D1519127

DESIGN SCIENCE COLLECTION

SERIES EDITOR
Arthur L. Loeb
Department of Visual and Environmental Studies,
Carpenter Center for the Visual Arts, Harvard University

Amy C. Edmonson
A Fuller Explanation: The Synergetic Geometry of R. Buckminster Fuller (1987)

Marjorie Senechal and George Fleck (editors)
Shaping Space: A Polyhedral Approach (1988)

Judith Wechshler (editor)
On Aesthetics in Science (1988)

Lois Swirnoff
Dimensional Color (1989)

Arthur L. Loeb
Space Structures: Their Harmony and Counterpoint (1991)

Arthur L. Loeb
Concepts and Images (1993)

Hugo F. Verheyen
Symmetry Orbits (1996)

Carl Bovill
Fractal Geometry in Architecture and Design (1996)

HUGO F. VERHEYEN

Symmetry Orbits

Birkhäuser
Boston • Basel • Berlin

Dr. Hugo F. Verheyen
Gloriantlaan 64-15B
B-2050 Antwerpen
Belgium

Library of Congress Cataloging-in-Publication Data

Verheyen, Hugo F. (Hugo François), 1949-
  Symmetry orbits / Hugo F. Verheyen.
    p.   cm. -- (Design science collection)
  Includes bibliographical references.
  ISBN 0-8176-3661-7 (acid-free). -- ISBN 3-7643-3661-7 (acid-free)
    1. Geometry. 2. Symmetry. 3. Symmetry groups. I. Title.
II. Series.
QA447.V47   1996
516--dc20   93-31298
                                                                                                CIP

Cover illustration: Model of tetrahedral compound of 4 cubes, with symmetry for group $A_4 \times I$, assembled by the author based on patterns by M.G. Fleurent, and photographed by H. Thiers-Konings.

Printed on acid-free paper.

© 1996 Birkhäuser Boston   *Birkhäuser*

Copyright is not claimed for works of U.S. Government employees.
All rights reserved. No part of this publication may be reproduced, stored in a retrieval system or transmitted, in any form or by any means, electronic, mechanical, photocopying, recording or otherwise, without prior permission of the copyright owner.

Permission to photocopy for internal or personal use, or the internal or personal use of specific clients, is granted by Birkhäuser Boston for libraries and other users registered with the Copyright Clearance Center (CCC), provided that the base fee of $6.00 per copy, plus $0.20 per page is paid directly to CCC, 222 Rosewood Drive, Danvers, MA 01923, U.S.A. Special requests should be addressed directly to Birkhäuser Boston, 675 Massachusetts Avenue, Cambridge, MA 02139, U.S.A.

ISBN 0-8176-3661-7
ISBN 3-7643-3661-7

Typeset by Sherman Typography, York, Pennsylvania
Printed and bound by Braun and Brumfield, Ann Arbor, Michigan
Printed in the United States of America

9 8 7 6 5 4 3 2 1

Series Editor's Foreword

In a broad sense design science is the grammar of a language of images rather than of words. Modern communication techniques enable us to transmit and reconstitute images without needing to know a specific verbal sequence language such as the Morse code or Hungarian. International traffic signs use international image symbols which are not specific to any particular verbal language. An image language differs from a verbal one in that the latter uses a linear string of symbols, whereas the former is multidimensional.

Architectural renderings commonly show projections onto three mutually perpendicular planes, or consist of cross sections at different altitudes capable of being stacked and representing different floor plans. Such renderings make it difficult to imagine buildings comprising ramps and other features which disguise the separation between floors, and consequently limit the creative process of the architect. Analogously, we tend to analyze natural structures as if nature had used similar stacked renderings, rather than, for instance, a system of packed spheres, with the result that we fail to perceive the system of organization determining the form of such structures.

Perception is a complex process. Our senses record; they are analogous to audio or video devices. We cannot, however, claim that such devices perceive. Perception involves more than meets the eye: it involves processing and organization of recorded data. When we name an object, we actually name a concept: such words *as octahedron, collage, tessellation, dome,* each designate a wide variety of objects sharing certain characteristics. When we devise ways of transforming an octahedron, or determine whether a given shape will tessellate the plane, we make use of these characteristics, which constitute the grammar of structure.

The Design Science Collection concerns itself with various aspects of this grammar. The basic parameters of structure such as symmetry, connectivity, stability, shape, color, size, recur throughout these volumes. Their interactions are complex; together they generate such concepts as Fuller's and Snelson's tensegrity, Lois Swirnoff's modulation of surface through color, self-reference in the work of M.C. Escher, or the synergetic stability of ganged unstable polyhedra. All these occupy some of the professionals concerned with the complexity of the space in which we live, and which we shape. The Design Science Collection is intended to inform a reasonably well educated but not highly specialized audience of these professional activities, and particularly to illustrate and to stimulate the interaction between the various disciplines involved in the exploration of our own three-dimensional, and, in some instances, more-dimensional spaces.

Hugo Verheyen addresses himself to a particularly abstract chapter in the grammar of space, namely to group theory. This chapter is full of technical terms which are difficult to visualize and accordingly for the visually oriented hard to remember. When group theory is taught, crystal symmetry is usually used as a physical illustration, but the crystallographic symmetry serves here to facilitate the learning of group theory rather than the other way around. Verheyen illustrates the concepts through the use of physical models, and accompanies these with instructions how to build these models.

In particular, Verheyen uses group theory for a systematic study of the nesting of regular, semi-regular and Archimedean solids. Such nesting involves the proper alignment of the symmetry elements of the nested solids, and the sharing of vertices or faces. The concept of subgroups thus acquires a physical reality and consequently a physical application. Interrelations between the simple Platonic solids and their complex derivatives are discovered and proven.

It is hoped that this volume will reveal to the reader new interrelations and transformations between a variety of polyhedra having variously tetrahedral, octahedral or icosahedral symmetries. And above all, the reader will be able to spend many happy hours building and exploring these elegant structures.

    Arthur L. Loeb
    Cambridge, Massachusetts

Contents

Series Editor's Foreword — v

Introduction — 1

PART I
Realization of Symmetry Groups

1 Groups of Isometries — 11
2 Symmetry Action — 67
3 Orbit Systems — 81

PART II
Compounds of Cubes

4 Classification of the Finite Compounds of Cubes — 95
5 Stability of Subcompounds — 161
6 Higher Descriptives — 195
7 Assembling Models — 201

Appendix: Historical Survey — 223
References — 233
Illustratory Contributions — 235

Introduction

In the realm of *symmetry* various disciplines meet, resulting in a combination of knowledge that extends the abilities of each separate discipline. Publications on symmetry-related subjects are now appearing from different angles of scientific research, and conferences are increasingly being held in places all over the world. Topics such as polyhedral shapes, domes, expandability, membranes, and lightweight structures in architecture all attract specialists of a wide variety; among these are architects, engineers, mathematicians, astronomers, crystallographers, chemists, biologists, physicists, artists, designers, industrialists, and more. This blending of information is occurring with a common goal: to achieve *a functional application of geometry*. In such a sense, this volume basically investigates the applicatory qualities of an *orbit* in group-theory, for Design Science.

1. A Term Inspired by Astronomy

Like many terms applied in mathematics, the term "orbit" was inspired by another discipline, from which it was borrowed and is used in a comparable sense. In astronomy, an orbit refers to the rotation of a certain object (A) about another object (B), like a planet orbiting the sun, or a satellite orbiting Earth. The orbiting object A describes a curve, often an ellipse, with B located at one of its focus points. While in orbit, A takes successive positions along the curve. From this idea of successive positions, the term "orbit" was introduced in group-theoretical algebra to refer to a set of mapped images of an element (A) combined with the elements of a *group*. This definition is situated within the topic of *group action on a set*, which shall be explained in the preliminary sections further on. For our scope, *groups of isometries* or *symmetry groups* are specifically selected, the elements of which are operators in three-dimensional Euclidean space. The chosen action of such operators then applied is the transformation of a body into a congruent body. Such an orbit of the body under a symmetry group is thus the set of transformed bodies under all the isometries of the group, which is an arrangement of congruent bodies having overall symmetry under the group. The parallel idea with the astronomical orbit is obvious: in either sense the object can be seen as describing a series of positions.

The expressions "active symmetry" and "symmetry action" thus stand for the operator group action of isometries in three-dimensional Euclidean space, as can be mathematically well understood.

2. Two Parts in this Volume

This book is obviously not written for theoretical mathematicians only. Yet, according to my experience, group theory may unfortunately quickly turn off our general readers. This is quite understandable, as sufficient knowledge of group-theoretical algebra is usually taken for granted by mathematical authors. Without some essential knowledge, the point is then likely to be missed. For example, in various articles and books, symmetry bodies are usually related with finite groups of isometries. However, I have found that the totality of facts and figures of symmetry groups can hardly be recovered from the literature, whereas such basic knowledge is absolutely vital. Hence, this volume is composed of two distinct parts, in which theory and practical applications are treated separately.

2.1. Preliminaries and Theory

In the first part, Chapter 1 may be considered as somewhat of a *textbook on isometry groups*. Here, the types of isometries are presented, and their finite groups are classified, discussed, and illustrated. The composition of each such group is explained, and all numerical data, such as angles between axes, dihedral angles between mirrors, etc., are listed in tables. With this information any reader should be able to obtain a good understanding of the groups of symmetry, which can be used as a further base for enjoying any publication on the subject.

Chapter 2 presents the very fundament of this study: the operator action of a group of isometries in three-dimensional Euclidean space, wherein the orbits are situated. For the benefit of the reader, the general algebraic approach is treated from the beginning, and gradually diverted into a specialized application to Euclidean geometry of three dimensions. The orbit becomes a "visual realization" of a symmetry group, in which the properties are transmitted, seen, and understood.

In Chapter 3, realizations of symmetry groups are discussed with examples. It is explained how an important application consists of a construction method for symmetry bodies and symmetric arrangements of these. Of such arrangements, special examples are those in which a solid with central symmetry is centrally positioned with respect to the group: these are known as *compounds*.

2.2. A Classification Method for Compounds

In the second part, the orbit method for constructing symmetric arrangements of solids is fully discussed for one particular example that has been selected for various reasons: a centrally positioned *hexahedron* or *cube*. The principle implies a full classification of the various types of such compounds. Because this classification is quite extensive, it may, therefore, be seen as a section of its own. The investigation of compounds can be situated in the discipline of *polytopology*, the study of the geometry of polyhedra and polytopes, with contributions of mathematicians during the last two centuries.

At the end of this book, a historical appendix is added, in which an explanation is presented of how some cube compounds with symmetry have been conceived by different persons through their own method of construction. Only a small number have been published in articles and books, and now the need of a serious mathematical approach has finally arisen.

Hence, the scope of Part II is a complete enumeration and classification of all cube compounds with finite symmetry. This is simply undertaken as the analysis of a particular realization example of the finite groups of isometries, discussed in Part I.

Claiming a discovery in scientific research has always been a delicate matter, and the history is full of disputes concerning that. It is generally accepted that the dedication of a discovery is established by the first official publication. I have also applied this standard in my historical review, without, however, neglecting the mention of unpublished contributions known to me.

I would prefer to use "introducing" when presenting the complete set of 30 symmetric compounds of cubes rather than emphasize the "discovery" of those that have not been published as yet. Single examples of compounds can be conceived by various methods. I may, however, state that the applied mathematical theory of orbits has allowed me consequently to derive the entire set of cube compound types in a logical way. As such, I may state that I have been *independently* able to classify them, of which 25 are *published* here for the first time, when deleting the single cube compound. These are (underscores indicate the addition of an illustration):

⋄ the 5 compounds with central freedom:

$$n \mid C_n \times I \mid E \times I$$
$$2n \mid D_n \times I \mid E \times I$$
$$12 \mid A_4 \times I \mid E \times I$$
$$24 \mid S_4 \times I \mid E \times I$$
$$60 \mid A_5 \times I \mid E \times I$$

⋄ 14 of the 15 compounds with rotational freedom:

$$\underline{2n \mid D_{4n} \times I \mid C_4 \times I}$$
$$\underline{2n \mid D_{3n} \times I \mid C_3 \times I}$$
$$\underline{2n \mid D_{2n} \times I \mid D_1 \times I}$$

$nA \mid D_n \times I \ / \ C_2 \times I$

$nB \mid D_n \times I \ / \ C_2 \times I$

$\underline{4 \mid A_4 \times I \ / \ C_3 \times I}$

$\underline{6 \mid A_4 \times I \ / \ C_2 \times I}$

$\underline{8 \mid S_4 \times I \ / \ C_3 \times I}$

$12 \mid S_4 \times I \ / \ D_1 \times I$

$\underline{12A \mid S_4 \times I \ / \ C_2 \times I}$

$12B \mid S_4 \times I \ / \ C_2 \times I$

$\underline{20 \mid A_5 \times I \ / \ C_3 \times I}$

$30A \mid A_5 \times I \ / \ C_2 \times I$

$30B \mid A_5 \times I \ / \ C_2 \times I$

◊ 6 of the 9 rigid compounds:

$n \mid D_{4n} \times I \ / \ D_4 \times I$

$\underline{n \mid D_{3n} \times I \ / \ D_3 \times I}$

$n \mid D_{2n} \times I \ / \ D_2 \times I$

$\underline{6 \mid S_4 \times I \ / \ D_2 \times I}$

$\underline{10 \mid A_5 \times I \ / \ D_3 \times I}$

$\underline{15 \mid A_5 \times I \ / \ D_2 \times I}$

This derivation principle can be easily applied to other Platonic, or even Uniform Polyhedra, and perhaps in the near future, I might challenge the compounds of *tetrahedra*, or even of dazzling *great inverted retrosnub icosidodecahedra* [5].

3. An Abundance of Illustrations

Throughout this book, the reader will note that much attention has been given to the illustrations: hand drawings, computer graphics, tables with data and figures, and photographs of models. I am convinced that a good understanding of three-dimensional symmetry properties can develop only in combination with the training of visualization capacities. All drawings and models of groups and compounds are, therefore, shown from different angles of view, and—with some occasional exceptions, when another view is justified—along different types of symmetry axes.

All of the illustrative materials were made exclusively for this book, except the compound models in Figs. 94, 102, 108, 131, and 135, of which pictures have been provided previously for other publications. Here, however, the photographs of the occurring models are all original and have been solely taken for publication in this book.

The models in Part I are made from plastic elements of a polyhedron "kit" that I designed some years ago. The die-cut triangles, squares, pentagons, and diamonds are equipped with foldable tabs and slots along the edges that snap together effortless but hold strongly. Dismantling afterwards is easy as the material is lasting, and the elements can be reused for later models. The pieces have been slightly modified to allow wooden rods to pass through their center, which meet in a wooden ball at the center of the solid.

As for the cardboard compound models in Part II—those with intersecting faces—a completely different preparatory design process was involved. First, for each compound, the core had to be determined, which is the exact shape of the inner convex polyhedron, isolated inside by the total number of square faces in the compound. Next, the stellation pattern of the core was calculated, by which the intersecting lines of each face with the remaining faces could be drawn within the boundary square. According to the compound involved, one, two, or three different face intersection patterns were to be hand-drawn, for which the visible parts on the outside had to be identified for the later model. Then, the edges of these elements were identified and numbered, according to where they had to be joined for the construction. Now, the patterns were ready for the final assembly process. The involved parts were printed onto flat cardboard sheets of different colors, and the edges were provided with tabs, cut out, and glued together along the fitting common edges, until a finished three-dimensional model had appeared.

4. Assembling Your Own Models

Despite the illustrative materials, which accompany the treatment of this subject, the desire for experiencing symmetries in a *three-dimensional* model may still prevail. Hence, in the last chapter of Part II, this opportunity will be offered, as the reader is invited to assembling cube compounds of his own choice in the form of cardboard models. Patterns and clear assembly instructions are provided for a substantial collection of models which were selected for their diversity and degree of complexity. There is no need to develop any special skill for the assembly of such a model: enthusiasm and lots of patience will finish the job.

5. Acknowledgments

Finally, I would like to express cordial thanks to a number of helpful friends; each have offered highly appreciated assistance in some way. Without their support, I could have never completed this project considering the variety of theoretical and practical tasks that had to be done.

Prof. Dr. Fred Van Oystaeyen (UIA University of Antwerp, FL, Belgium), has truly encouraged me to provide a more elaborate section on group theory than I originally had planned. At first I was going to apply definitions and properties directly to the realm of symmetry. One afternoon, in his large office at the university, I brought up the idea of applying part of the orbit theory specifically to constructions under symmetry groups. Fred found this to be a suitable approach to this subject, but made it clear that it should be done in the proper mathematical way, that is, from a well-established theoretical base. Fred also made invaluable suggestions for the section on action and orbits.

Gilbert M. Fleurent, O.Pr. (Abbey of Averbode, FL, Belgium) provided special computer graphics of symmetry groups seen along the different kinds of axes. He also was specifically involved with the evolution of the models of cube compounds for Part II. Working daily for more than a year, Gilbert calculated, drew, and analyzed the patterns in order to provide the proper assembly instructions for the three-dimensional models—with an exception of three such patterns which were similarly worked out by his correspondence friend Peter Messer, M.D. (Mequon, WI).

Although an excellent model maker, Gilbert could not spend more time assembling his own models, since he was promoted to chief librarian of the Abbey. I then assembled five models from these patterns. The remaining set—among which were the most intricate ones—was offered to be finished by Magnus J. Wenninger (St. John's Abbey, MN), who is a master model maker. These patterns were sent to Minnesota, where Magnus would spend a few thousand hours of careful assembly.

Then came the photographic question. I definitely wanted to make sure that pictures would show models from different kinds of axes. Therefore, I wanted to photograph the models personally with my own camera equipment. Those in my own possession were no problem, but the American set would prove too much of a risk to have them shipped to Europe. So I went to visit Magnus once again, and took pictures of all of his models there. After my return, most of the photos proved to be satisfactory for publication, with the exception of a small set that was overlighted on one side (the pictures had been taken outside, and weather conditions were not great at St. John's). That set was reproduced later by the Abbey's photographer, Don Bruno (St. Cloud, MN) and sent to us. My personal set was partly photographed at my parents' house in Schilde, near Antwerp, and partly at the home of a photographer and a good friend of mine, Hedwig Thiers-Konings (Brasschaat, FL, Belgium), with professional equipment.

PART I
Realization of Symmetry Groups

Chapter 1

Groups of Isometries

The purpose of this chapter is to offer a compendium of things worth knowing about isometries, in general, and especially about their finite groups. Often in the literature, this knowledge is taken for granted, whereas in practice the composition, angles between axes, and other data are seldom listed. Most proofs of properties belong to elementary geometry and can be found here and there in such treatments, as, e.g., with Coxeter [3, 4]. Definitions and a number of properties will be merely mentioned here, but a full list of finite groups of isometries will be presented, together with numerical and geometric data, and all provided with clear illustrations.

1. Classification of Isometries

The three-dimensional Euclidean space, denoted by E^3, has been provided with the definition of *distance*, a standard subject of study in *topology*. A transformation in E^3, which preserves the distance of any two points, is known as an *isometry*. As a result, a subset of E^3 is transformed into a congruent subset by any isometry. Some authors prefer to define an isometry as a finite product of reflections. Whichever definition is chosen, the other will consequently become a property. Personally, I prefer an isometry being defined as a finite product of reflections, that is, of course, after the definition of reflection has been outlined:

> let ω be a plane in E^3, P' denote the projection of a point $P \in E^3$ in ω. A transformation T:

$$T: E^3 \longrightarrow E^3$$
$$P \longmapsto Q$$

such that $|PP'| = |P'Q|$, where P and Q lie on opposite sides of ω on line PP', is a *reflection* in E^3, and ω is generally known as the *mirror* of the reflection.

1.1. The Types of Isometries

It can be easily proved that any isometry is a product of at least four reflections and is one of the following six types:

1. *Reflection*

> Product number of reflections: 1 (Fig. 1)
> Points of invariancy: plane (the mirror)

2. *Rotation*

> Product number of reflections: 2
> Points of invariancy: line
> The mirrors are intersecting along the line of invariancy, which is called the *axis* of the rotation (Fig. 2). If α is the

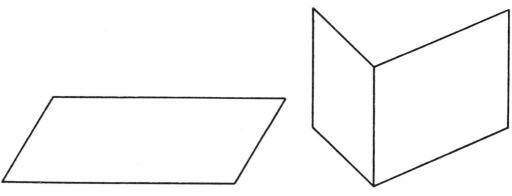

Figure 1. Reflection

Figure 2. Rotation

sharp (or right) dihedral angle of the mirrors, the *angle* of the rotation is 2α in one sense.

Two special cases are observed for which the product of the two reflections is commutative:

a. the reflections are identical

The product is E, the *identity*, and the dihedral angle between the mirrors is 0°, turning the identity into a rotation through 0°. All points are points of invariancy.

b. The mirrors are perpendicular

This product is a rotation through 180°, also called a *half-turn*.

3. *Translation*

Product number of reflections: 2

Points of invariancy: none.

The mirrors are parallel (Fig. 3). This type can be considered as a rotation whose axis is at infinity. If the distance between both mirrors is d, the translation is along a direction perpendicular to both planes and over a distance $2d$ in one sense.

The identity is also a special case of a translation, namely, when the parallel planes coincide. This is a translation over a distance 0.

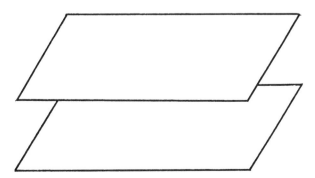

Figure 3. Translation

4. Rotatory Reflection

Product number of reflections: 3
Points of invariancy: point
Two mirrors are intersecting and the third is perpendicular to both, resulting in a commutative product of a rotation and a reflection (Fig. 4). The point 0, common to the three mirrors, is the point of invariancy.

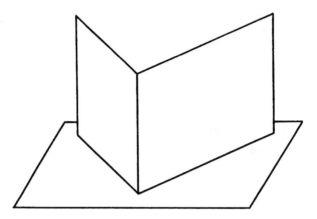

Figure 4. Rotatory reflection

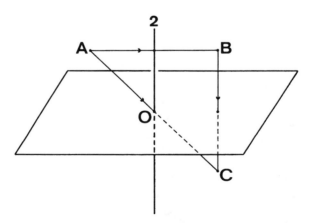

Figure 5. Central inversion

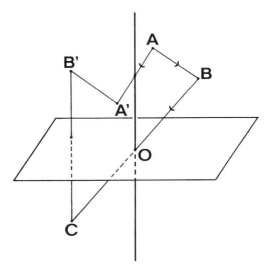

Figure 6. A rotatory reflection is a rotatory inversion. (a) A is mapped onto B (rotation) and B onto C (central inversion). (b) A is mapped onto A' (half-turn), A' onto B' (rotation), and B' onto C (reflection)

> A special case is observed:
> If the two mirrors are perpendicular, the rotation is a half-turn, and the rotatory reflection is called a *central inversion* about the point 0 (Fig. 5). The three mirrors are now mutually perpendicular, and the product of the three reflections is entirely commutative.
> Because a reflection is also the product of a half-turn and a central inversion, any rotatory reflection can also be considered as a *rotatory inversion*, where the angle of the new rotation has been increased by 180° (Fig. 6).

5. *Glide Reflection*

> Product number of reflections: 3
> Points of invariancy: none
> Two mirrors are parallel and the third is perpendicular to both (Fig. 7), resulting in a commutative product of a translation and a reflection.

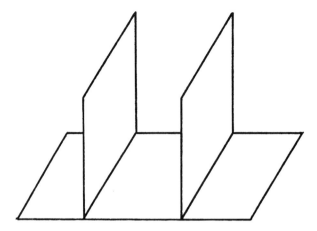

Figure 7. Glide reflection

6. *Glide Rotation or Twist*

   Product number of reflections: 4
   Points of invariancy: none
   Two mirrors are parallel, and the two others intersect and are perpendicular to both (Fig. 8), resulting in a commutative product of a translation and a rotation.

According to this classification, three types of isometries have points of invariancy: reflection, rotation, and rotatory reflection. The three remaining types have no points of invariancy and are often also referred to as *displacements:* translation, glide reflection, and twist.

1.2. Direct and Opposite Isometries

A distinction is made between *even* and *odd* products of reflections: An even product is called a *direct* isometry, and an odd product an *opposite* isometry.

Table 1 lists seven types of isometries according to this distinction.

An interesting consequence is formulated by the next general property: **Every direct isometry is the product of two half-turns, and every opposite isometry is the product of a half-turn**

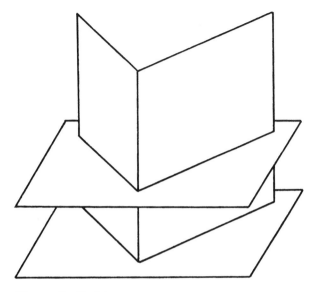

Figure 8. Twist

Table 1

Isometry	Type	Points of Invariancy
Identity	Direct	Space
Reflection	Opposite	Plane
Rotation	Direct	Line
Translation	Direct	—
Rotatory reflection	Opposite	Point
Glide reflection	Opposite	—
Twist	Direct	—

Table 2

Position	Type
Coincidental	Identity
Intersecting	Rotation
Parallel	Translation
Skew	Twist

Table 3

Position	Type
Coplanar	Reflection
Intersecting	Rotatory reflection
Parallel	Glide reflection

and a reflection. Tables 2 and 3 illustrate the distinct types of direct and opposite isometries, respectively. In Table 2, the relative positions of the axes of the two half-turns are discussed, and in Table 3 those of the axis and the mirror.

2. Groups of Isometries

The product of two isometries is again a finite product of reflections. A product of two direct or opposite isometries is always direct, and a product of a direct and an opposite isometry is opposite.

Moreover, every isometry has an inverse:

$$(T_1 \cdot T_2 \cdots T_n) \cdot (T_n \cdots T_2 \cdot T_1) = E$$

Hence, the total set **I** of isometries in E^3 is a *group*. The direct isometries in **I** consist of a subgroup of index 2.* Let S be an opposite isometry. Any opposite isometry T can then be expressed as

$$T = (S \cdot S^{-1}) \cdot T = S \cdot (S^{-1} \cdot T)$$

whereas $S^{-1} \cdot T$ is direct.

Subgroups occurring in **I** may be either *discrete* or *finite*, e.g., $\{E, T\}$, where T denotes a reflection, is a group of order 2 in **I**. If T denotes a translation, the group generated by T is a discrete group of translations.

A group of isometries is then obviously either composed of direct isometries only or contains such a subgroup of index 2, whereas the coset contains all the opposite isometries.

The *finite groups of isometries* will receive special attention later as it needs an extensive description.

3. The Finite Groups of Isometries

All finite groups of isometries have at least one point of invariancy [3] and, hence, do not contain displacements (see Section 1.1). Such as group is then either composed of rotations only or contains such a subgroup of index 2 whose coset is composed of reflections and/or rotatory inversions.

3.1. The Finite Groups of Rotations

These groups are classified in isomorphic families of five categories: cyclic, dihedral, tetrahedral, octahedral, and icosahedral. The cyclic and dihedral categories consist of an infinite number and are denoted respectively by C_n and D_n ($n \geq 1$). C_n contains n rotations (including the identity) and D_n contains $2n$ rotations. The tetrahedral group is isomorphic with A_4, the alternating group of 4 elements, as can be easily understood by identifying the 4 elements each with a face of a tetrahedron and permuting them by the 12 rotations of the group. By embedding the group in the octahedral or the icosahedral group and having the tetrahedron's faces permuted by the rotations of such a group, it is similarly found that these groups are isomorphic with S_4 and A_5, respectively, S_4 being the permutation group of four elements and A_5 the alternating group of five elements. Table 4 illustrates these groups.

3.2. The Finite Extended Groups

The expression "extended" group was introduced by Klein [7] and refers to a higher group **F**, containing a group **G** of rotations

Table 4

Group	Order
C_n	n
D_n	$2n$
A_4	12
S_4	24
A_5	60

* Definitions of *index* and *coset* can be found in Chapter 2, Section 1.1.

as a subgroup of index 2, whose coset contains all the opposite isometries. There are two ways to extend a group **G** of rotations into a group **F**:

1. *Groups Containing I*

Here, the coset is obtained by multiplying all the rotations of **G** with I, the central inversion. The symbol used is $\mathbf{G} \times I$:

$$\mathbf{G} \times I = \mathbf{G} \cup \mathbf{G} \cdot I$$

Such a group contains I as an element because $E \in \mathbf{G}$ and $E \cdot I = I$. Table 5 lists all of these extensions.

2. *"Mixed" Groups*

This term was introduced by Coxeter [3] and refers to a derivation from a higher group **G**′ of rotations, containing **G** as a subgroup of index 2, that is if such a group **G**′ exists. If the coset of **G** in **G**′ is multiplied by I, clearly a new group is obtained when this set of opposite isometries is united with **G**. Such a mixed group is denoted by **G**′**G**:

$$\mathbf{G}'\mathbf{G} = \mathbf{G} \cup (\mathbf{G}' - \mathbf{G}) \cdot I$$

This extended group does not contain **I** as an element as $E \notin \mathbf{G}' - \mathbf{G}$. Table 6 lists all groups **G**′ of rotations that contain a subgroup **G** of index 2.

As a result, the mixed groups are now classified in Table 7.

In the following section, all finite groups of isometries will be separately described and analyzed.

4. Analysis of the Finite Groups of Rotations

For each type of group, the following data will be provided:

◊ a table containing the symbol for the group, the number of axes determined by cyclic subgroups of a certain maximal order, and the total number of axes of the group

Chapter 1 Groups of Isometries

Table 5

Extended Group	Order
$C_n \times I$	$2n$
$D_n \times I$	$4n$
$A_4 \times I$	24
$S_4 \times I$	48
$A_5 \times I$	120

Table 6

Group	Subgroups of Index 2
C_{2n}	C_n
D_n	C_n
D_{2n}	D_n, C_{2n}
S_4	A_4

Table 7

Mixed Group	Order
$C_{2n}C_n$	$2n$
$D_n C_n$	$2n$
$D_{2n}D_n$	$4n$
$S_4 A_4$	24

Table 8

Symbol	n-Fold
C_n	1

⋄ a table with all the possible angles between axes
⋄ coplanarity of at least three distinct axes
⋄ illustrations of the set of axes
⋄ a discussion of occurring subgroups with illustrations when appropriate

For the cyclic and dihedral groups, illustrations for examples will be limited to $n = 5$. Higher groups are illustrated by pictures of models, based on certain symmetric polyhedra, and photographed closely along each of the different kinds of axes. The face associated with a certain axis indicates the order of the cyclic subgroup of rotations about that axis: a square or diamond for order 2, and $\{n\}$ (a regular n-gon) for order n.

4.1. Cyclic

If n is prime, the group has no proper subgroups; if n is not and f is a factor of n, C_f is a subgroup of order f and index n/f in C_n. If n is even, a half-turn is contained in C_n.

In the examples, a polygon $\{n\}$ about the axis has been added which is left invariant by the rotations of C_n.

1. C_1

   Order 1: E (the identity)

2. C_2

   Order 2: E
           1 half-turn   (Fig. 9)

3. C_3

   Order 3: E
           2 threefold rotations   (Fig. 10)

4. C_4

   Order 4: E
           1 half-turn
           2 fourfold rotations   (Fig. 11)
   C_2 is a subgroup of index 2.

5. C_5

   Order 5: E
           4 fivefold rotations   (Fig. 12)

4.2. Dihedral

The dihedral groups D_n are obtained by adding n half-turns to C_n; axes of the half-turns are coplanar, and perpendicular to the n-fold axis. These twofold axes connect opposite vertices of an imaginary polygon $\{2n\}$ about the n-fold axis. If n is even, the axes occur in perpendicular pairs, and if n is odd, none of the axes are perpendicular to one another. Table 9 lists the rotational axes.

C_n is a subgroup of order n and index 2 in D_n. As an analogue to 4.1, n may be prime, in which case there are no other

Chapter 1 Groups of Isometries

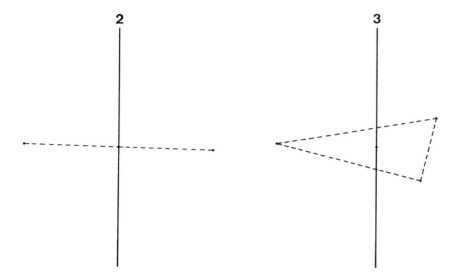

Figure 9. C_2

Figure 10. C_3

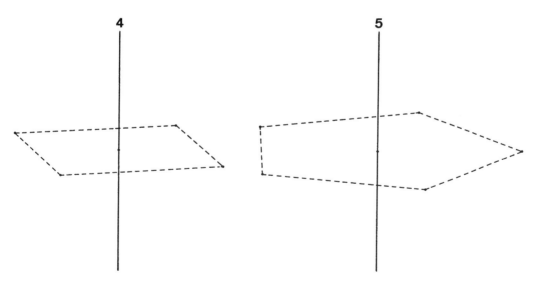

Figure 11. C_4

Figure 12. C_5

Table 9. Dihedral axes.

Symbol	2-Fold	n-Fold	Total
D_n	n	1	$n+1$

Table 10. Dihedral angles.

Axes		Angle	Supplement
2	2	$k/n \cdot 180°$ $(1 \leq k \leq n/2)$	
2	n	90°	90°

proper subgroups. If n is not and f is a factor of n, C_f is a cyclic subgroup of order f and index $2n/f$, and D_f is a dihedral subgroup of order $2f$ and index n/f.

In the examples for D_n, an imaginary polygon $\{2n\}$ has been added to the illustrations.

1. D_1

    Order 2: E
            1 half-turn
   $D_1 \cong C_2$ (see Fig. 9)

2. D_2

    Order 4: E
            3 half-turns   (Fig. 13)

3. D_3

    Order 6: E
            3 half-turns
            2 threefold rotations ($\pm 60°$)   (Figs. 14–15)

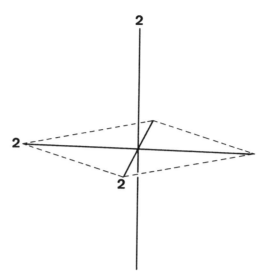

Figure 13. D_2

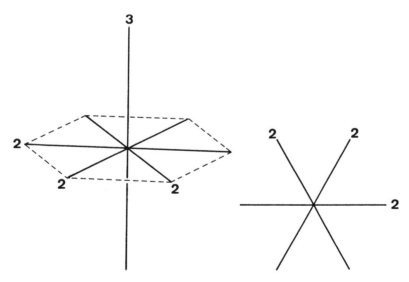

Figure 14. D_3

Figure 15. Coplanar axes of D_3, perpendicular to the threefold axis ($1\times$)

26　　　　　　　　　　　　　　　　　　　　　　　　Part I　Realization of Symmetry Groups

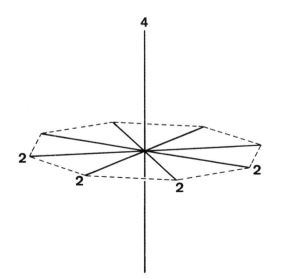

Figure 16. D_4

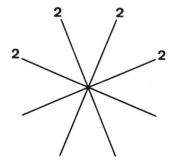

Figure 17. Coplanar axes of D_4, perpendicular to the fourfold axis ($1\times$)

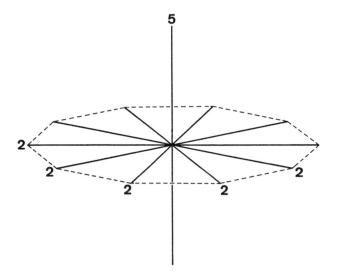

Figure 18. D_5

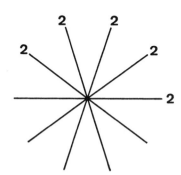

Figure 19. Coplanar axes of D_5, perpendicular to the fivefold axis ($1\times$)

Chapter 1   Groups of Isometries

4. D_4

Order 8: E
5 half-turns
2 fourfold rotations ($\pm 45°$)   (Figs. 16–17)
One of the half-turns is about the fourfold axis. D_2 is a dihedral subgroup of index 2.

5. D_5

Order 10: E
5 half-turns
4 fivefold rotations ($\pm 72°$, $\pm 144°$)
(Figs. 18–19)

4.3. Tetrahedral

A_4

Order 12: E
3 half-turns
8 threefold rotations

The seven axes of A_4 are illustrated on a *cuboctahedron*, an Archimedean solid composed of six squares and eight triangles (see Fig. 20). The squares indicate the twofold axes and the triangles the threefold axes. Figure 21 shows coplanar axes.

D_2 is a dihedral subgroup of index 3 and is illustrated in Fig. 22. The remaining axes are the four threefold axes, shown in Fig. 23.

Table 11. Tetrahedral axes.

Symbol	2-Fold	3-Fold	Total
A_4	3	4	7

Table 12. Tetrahedral angles.

Axes		Angle	Supplement
2	2	90°	90°
2	3	54°44′08″.20	125°15′51″.80
3	3	70°31′43″.60	109°28′16″.40

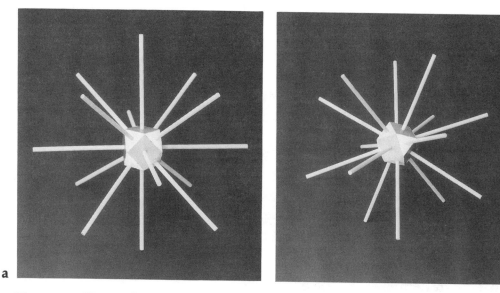

Figure 20. Views of A_4: (a) with a vertical twofold axis; (b) with a vertical threefold axis

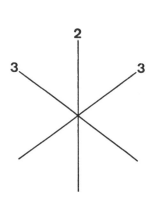

Figure 21. Coplanar axes of A_4 (6×)

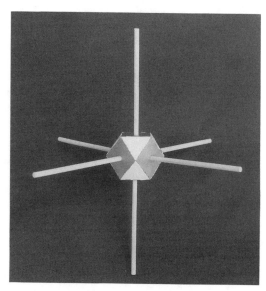

Figure 22. $D_2 \subset A_4$

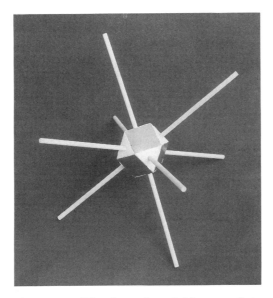

Figure 23. The four threefold axes of A_4

4.4. Octahedral

S_4

Order 24: E
9 half-turns
8 threefold rotations
6 fourfold rotations

Three of the nine half-turns are about the fourfold axes.

The 13 axes of S_4 are illustrated on a *rhombicuboctahedron*, an Archimedean solid with 8 triangles and (6 + 12) squares: 6 squares (dark) are about fourfold axes and 12 (light) about twofold axes, all in opposite pairs (Fig. 24).

This group contains A_4 as a subgroup of index 2, and, hence, the set of axes of A_4 is a subset of those of S_4. The three half-turns of A_4 are those belonging to the three cyclic sub-

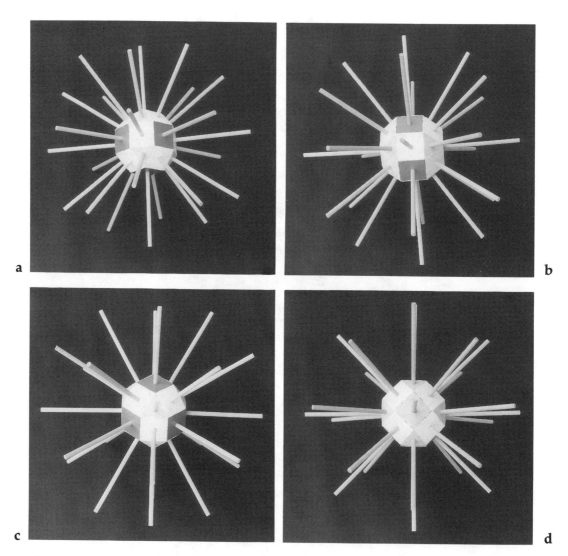

Figure 24. Views of S_4: (a) with a vertical fourfold axis; (b) along a twofold axis; (c) along a threefold axis; (d) along a fourfold axis

Chapter 1 Groups of Isometries

Table 13. Octahedral axes.

Symbol	2-Fold	3-Fold	4-Fold	Total
S_4	6	4	3	13

Table 14. Octahedral angles.

Axes		Angle	Supplement
2	2	60°	120°
		90°	90°
2	3	35°15'51".80	144°44'08".20
		90°	90°
2	4	45°	135°
		90°	90°
3	3	70°31'43".60	109°28'16".40
3	4	54°44'08".20	125°15'51".80
4	4	90°	90°

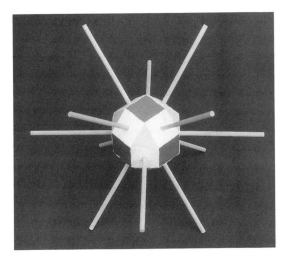

Figure 25. The six twofold axes of S_4

groups C_4 in S_4, and, hence, the three twofold axes of A_4 are identical with the fourfold axes of S_4. Six extra twofold axes appear in S_4 (shown in Fig. 25). Table 14 lists the axis angles and Fig. 26 illustrates their coplanarity.

The following dihedral subgroups are part of S_4: D_2 of order 4 and index 6 (Fig. 27), D_3 of order 6 and index 4 (Fig. 28), and D_4 of order 8 and index 3 (Fig. 29).

Because two different kinds of twofold axes exist, namely, those that belong to maximal cyclic subgroups of order 2 or 4, two kinds of subgroups D_2 occur in S_4. These can be indicated by the types of three mutually perpendicular axes, namely, 2-2-4 and 4-4-4.

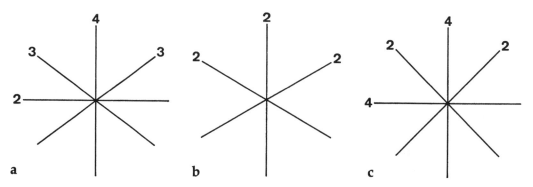

Figure 26. Coplanar axes of S_4: (a) perpendicular to a twofold axis ($6\times$); (b) perpendicular to a threefold axis ($4\times$); (c) perpendicular to a fourfold axis ($3\times$)

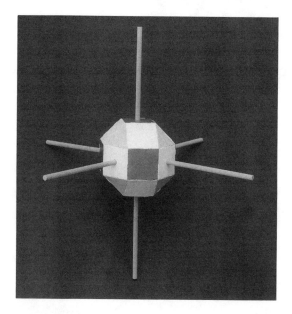

Figure 27. $D_2 \subset S_4$

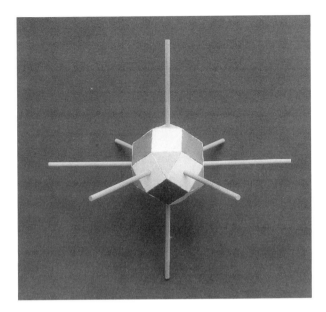

Figure 28. $D_3 \subset S_4$

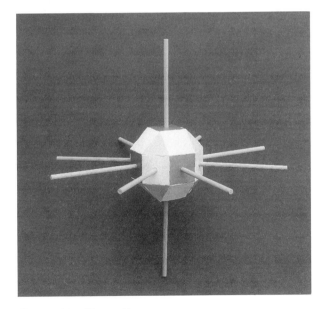

Figure 29. $D_4 \subset S_4$

4.5. Icosahedral

A_5

Order 60: E
15 half-turns
20 threefold rotations
24 fivefold rotations

The threefold and fivefold axes are first illustrated in an *icosidodecahedron*, an Archimedean solid with 20 triangles and 12 pentagons, each in opposite pairs about one axis (Figs. 30–31). The twofold axes are illustrated in its *dual* solid, the *rhombic triacontahedron*, composed of 30 golden rhombi (Fig. 32) [11]. Also, three mutually perpendicular twofold axes are emphasized, indicating a dihedral subgroup D_2 of order 4 and index 15 in A_5.

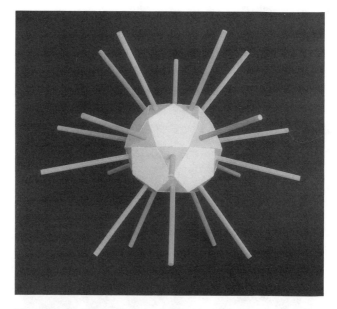

Figure 30. The 10 threefold axes of A_5

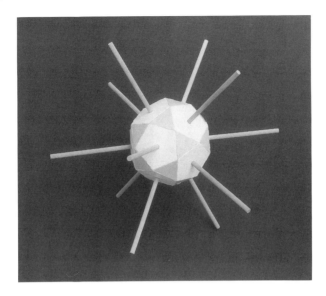

Figure 31. The 6 fivefold axes of A_5

Table 15. Icosahedral axes.

Symbol	2-Fold	3-Fold	5-Fold	Total
A_5	15	10	6	31

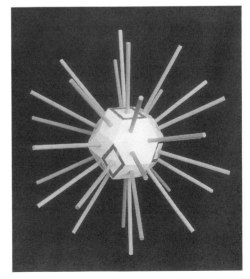

Figure 32. The 15 twofold axes of A_5, consisting of 5 sets of 3 mutually perpendicular axes each, illustrating also $D_2 \subset A_5$

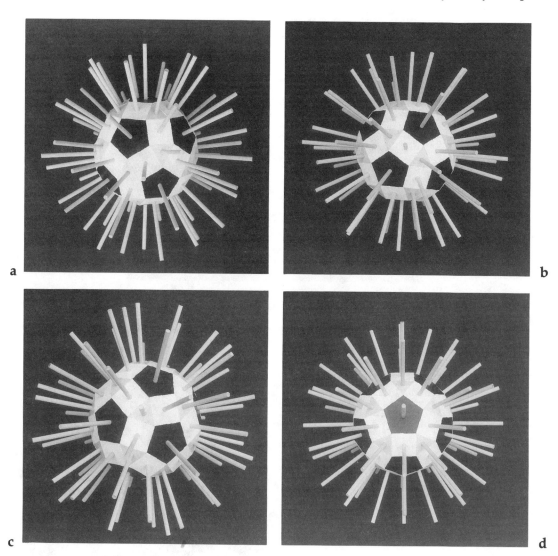

Figure 33. Views of A_5: (a) with a vertical fivefold axis; (b) along a twofold axis; (c) along a threefold axis; (d) along a fivefold axis.

Next, the totality of 31 axes is illustrated in a *rhombicosidodecahedron*, an Archimedean solid composed of 30 squares, 20 triangles, and 12 pentagons (Fig. 33), with a view to the inside of the model showing the center of the group, the point of invariancy O (Fig. 34).

Four threefold axes belong to a subgroup A_4 of order 12 and index 5 (Fig. 35).

Finally, the other dihedral subgroups are also illustrated on a rhombicosidodecahedron: D_3 of order 6 and index 10 (Fig. 36) and D_5 of order 10 and index 6 (Fig. 37). Table 16 lists the axes and Fig. 38 shows their coplanarity.

Figure 34. The center of A_5 is illustrated by a wooden sphere, in which the 31 axes meet

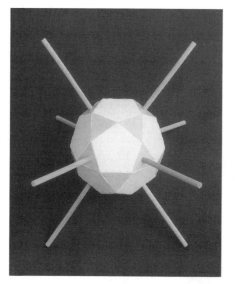

Figure 35. 4 threefold axes of $A_3 \subset A_5$

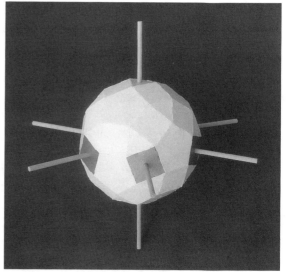

Figure 36. $D_3 \subset A_5$

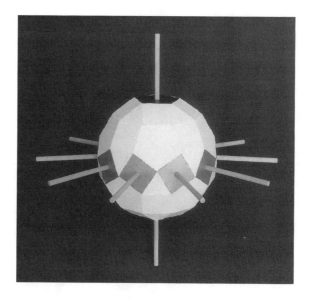

Figure 37. $D_5 \subset A_5$

Chapter 1 Groups of Isometries

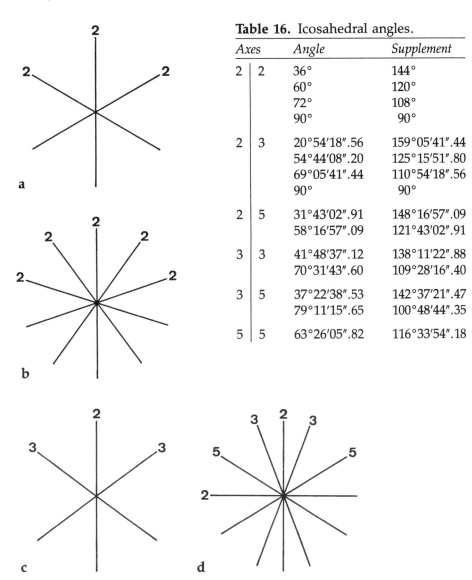

Table 16. Icosahedral angles.

Axes		Angle	Supplement
2	2	36°	144°
		60°	120°
		72°	108°
		90°	90°
2	3	20°54′18″.56	159°05′41″.44
		54°44′08″.20	125°15′51″.80
		69°05′41″.44	110°54′18″.56
		90°	90°
2	5	31°43′02″.91	148°16′57″.09
		58°16′57″.09	121°43′02″.91
3	3	41°48′37″.12	138°11′22″.88
		70°31′43″.60	109°28′16″.40
3	5	37°22′38″.53	142°37′21″.47
		79°11′15″.65	100°48′44″.35
5	5	63°26′05″.82	116°33′54″.18

Figure 38. Coplanar axes of A_5: (a) 30×; (b) perpendicular to a twofold axis (15×); (c) perpendicular to a threefold axis (10×); (d) perpendicular to a fivefold axis (6×)

5. Analysis of the Extended Groups

All possible extensions of finite groups of rotations will now be discussed. In both methods of construction (see Section 3.2), the added opposite isometries are obtained by the multiplication of a set of rotations (either **G** or **G'** − **G**) with I, the central inversion. This will result in the addition of an equal number of opposite isometries, altogether consisting of the coset of the subgroup of rotations in the extension. If the construction set contains half-turns, the associated opposite isometries will be reflections, and all the other products will result in rotatory inversions, including I—if it occurs.

For each type of group, the following data will be provided:

◊ a table containing the symbol for the group and its order, the identity, the number of rotations,* reflections, rotatory inversions,† and the central inversion—if it occurs
◊ a summary of all possible dihedral angles between mirrors in the group
◊ drawn illustrations of the axes and mirrors in the group. For the groups higher than cyclic and dihedral, i.e., the tetra-, octa-, and icosahedral groups, the intersections of the axes and mirrors with spherical surfaces will be illustrated by computer graphics viewed along each different type of axis
◊ a table containing the different kinds of mirrors with respect to their contents of axes, and illustrations of these.

5.1. Cyclic

C_n can be given three extensions:

1. $C_n \times I$
   The coset is $C_n \cdot I$. Only when n is even does C_n contain a half-turn. Hence, the coset is composed of I and

---

* Distinct from the identity.
† Distinct from reflections or the central inversion.

Chapter 1 Groups of Isometries

($n - 1$) rotatory inversions when n is odd, and one rotatory inversion becomes a reflection when n is even.

2. $C_{2n}C_n$

   The coset is obtained by subtracting C_n from C_{2n} and multiplying this set with I. When n is even, the half-turn in C_n is subtracted from C_{2n}, and when n is odd, it is not. Hence, the coset is composed of n rotatory inversions when n is even, and one of these becomes a reflection when n is odd.

3. D_nC_n

   When C_n is subtracted from D_n, the n added half-turns remain (see Section 4.2). The coset is then composed of n reflections, all intersecting along the n-fold axis. The dihedral angles of the mirrors are

$$(k/n)\,180° \quad (1 \leq k < n/2)$$

Specifications follow in Tables 17 and 18.

Table 17

Symbol	n^a	Id.	Rotat.	Reflections	Rot.-Inv.	Centr. Inv.	Total
$C_n \times I$	Even	1	$n-1$	1	$n-2$	1	$2n$
	Odd	1	$n-1$	—	$n-1$	1	$2n$
$C_{2n}C_n$	Even	1	$n-1$	—	n	—	$2n$
	Odd	1	$n-1$	1	$n-1$	—	$2n$
D_nC_n		1	$n-1$	n	—	—	$2n$

[a] $n \geq 2$ (for $n = 1$: see Example 1).

Table 18

Group	n	Mirrors	n-Fold
$C_n \times I$	Even	1	1
	Odd	—	—
$C_{2n}C_n$	Even	—	—
	Odd	1	1
D_nC_n		n	1

In the following examples a polygon $\{n\}$ about the n-fold axis will illustrate the transformations. In $C_n \times I$, C_n transforms $\{n\}$ into itself, whereas $C_n \cdot I$ maps it onto an inverse image, which is also a reflected image when n is even. For the mixed groups, $\{n\}$ is inscribed in a polygon $\{2n\}$. The remaining vertices of $\{2n\}$ determine those of a rotated image $\{n\}'$ of $\{n\}$, under $C_{2n} - C_n$. Again, C_n transforms $\{n\}$ into itself, whereas $(C_{2n} - C_n) \cdot I$ maps it onto an inverse image of $\{n\}'$, which is also a reflected image of $\{n\}$ when n is odd:

1. Extensions of C_1
   $C_1 \cong \{E\}$ (see Section 4.1)
   a. $C_1 \times I$ shall be denoted by $E \times I$
      Order 2: E
      I
   b. $C_2 C_1$
      Order 2: E
             1 reflection
      $C_2 - C_1$ is composed of a single half-turn; hence, the coset is composed of a single reflection.
   c. $D_1 C_1$
      $D_1 \cong C_2$ (see Section 4.2), hence this case is analogous with case b.
2. Extensions of C_2
   a. $C_2 \times I$
      Order 4: E
             I
             1 rotation
             1 reflection   (Fig. 39)
   b. $C_4 C_2$
      Order 4: E
             1 rotation
             2 rotatory inversions   (Fig. 40)
   c. $D_2 C_2$
      Order 4: E
             1 rotation
             2 reflections   (Fig. 41)
      Dihedral angles of the mirrors: 90°

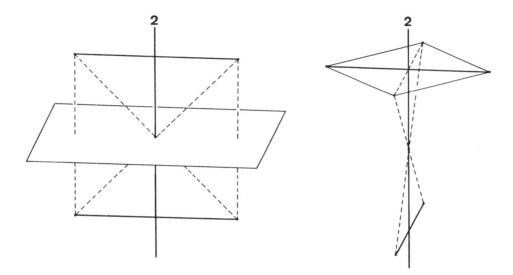

Figure 39. $C_2 \times I$

Figure 40. $C_4 C_2$

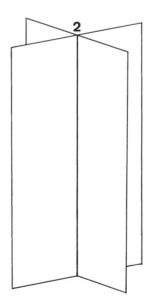

Figure 41. $D_2 C_2$

3. Extensions of C_3
   a. $C_3 \times I$
      Order 6: E
      I
      2 rotations
      2 rotatory inversions   (Fig. 42)
   b. $C_6 C_3$
      Order 6: E
      2 rotations
      1 reflection
      2 rotatory inversions   (Fig. 43)
   c. $D_3 C_3$
      Order 6: E
      2 rotations
      3 reflections   (Fig. 44)
      Dihedral angles of the mirrors: 60°

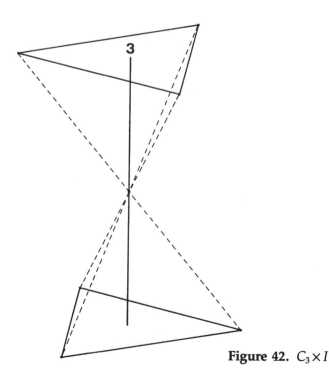

Figure 42. $C_3 \times I$

Chapter 1 Groups of Isometries

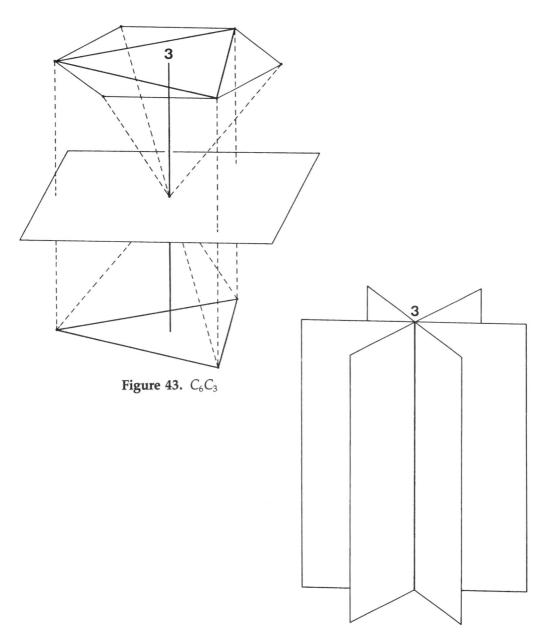

Figure 43. C_6C_3

Figure 44. D_3C_3

4. Extensions of C_4
   a. $C_4 \times I$
      Order 8: E
      I
      3 rotations
      1 reflection
      2 rotatory inversions   (Fig. 45)
   b. $C_8 C_4$
      Order 8: E
      3 rotations
      4 rotatory inversions   (Fig. 46)
   c. $D_4 C_4$
      Order 8: E
      3 rotations
      4 reflections   (Fig. 47)
      Dihedral angles of the mirrors: 45°, 90°

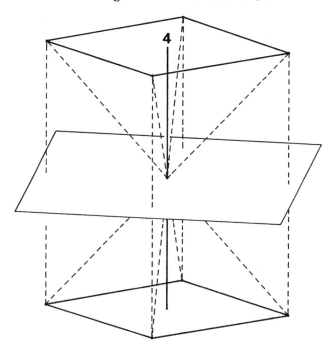

Figure 45. $C_4 \times I$

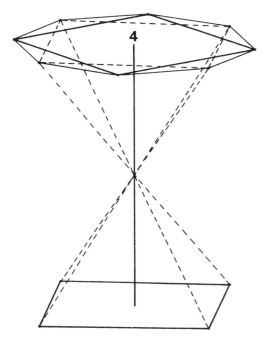

Figure 46. C_8C_4

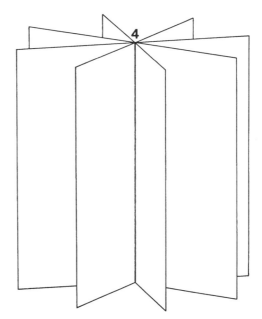

Figure 47. D_4C_4

5. Extensions of C_5
   a. $C_5 \times I$
      Order 10: E
      I
      4 rotations
      4 rotatory inversions   (Fig. 48)
   b. $C_{10}C_5$
      Order 10: E
      4 rotations
      1 reflection
      4 rotatory inversions   (Fig. 49)
   c. D_5C_5
      Order 10: E
      4 rotations
      5 reflections   (Fig. 50)
      Dihedral angles of the mirrors: 36°, 72°

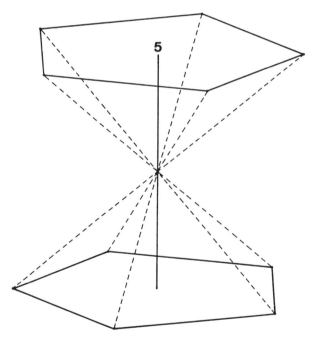

Figure 48. $C_5 \times I$

Chapter 1 Groups of Isometries

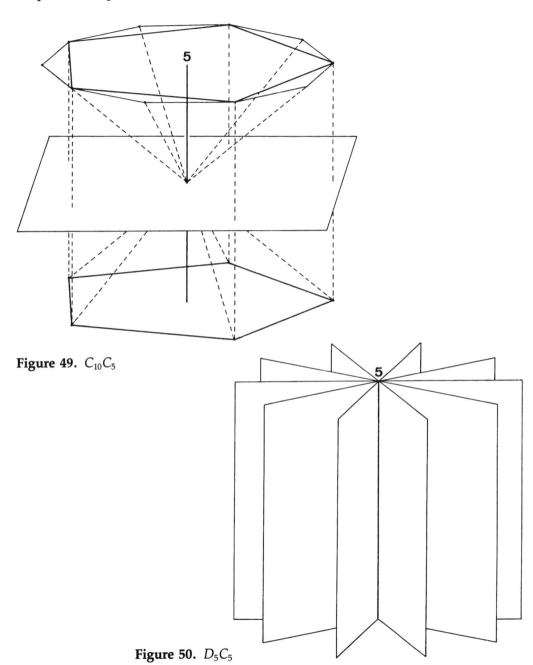

Figure 49. $C_{10}C_5$

Figure 50. D_5C_5

5.2. Dihedral

Because $D_n - C_n$ contains only half-turns, each of both extensions of D_n is the sum of an extension of C_n and n reflections, whose mirrors all intersect along the n-fold axis:

a. $D_n \times I$

If n is even, the axes of the n half-turns of D_n occur in perpendicular pairs (see Section 4.2). Hence, each one belongs to a mirror added to $C_n \times I$. Also, there is an extra reflection whose mirror contains all these axes (see Section 5.1). If n is odd, there are no such perpendicular axes, and they now become bisectors of the n mirrors, which are all alike and contain only the n-fold axis.

b. $D_{2n}D_n$

Because now D_n is subtracted from D_{2n}, the opposite situation of case a occurs when "even" and "odd" are interchanged.

Dihedral mirror angles: $90°$, $(k/n)180°$ $(1 \leq k \leq n/2)$

Specifications are shown in Tables 19 and 20.

In the examples, the imaginary polygon $\{2n\}$ (see Section 4.2) is visible in the illustrations:

1. Extensions of D_1
   $D_1 \cong C_2$ (see Section 4.2)
   a. $D_1 \times I \cong C_2 \times I$
   b. $D_2 D_1 \cong D_2 C_2$
2. Extensions of D_2
   a. $D_2 \times I$
   Order 8: E
   I
   3 rotations
   3 reflections   (Fig. 51)

   Here we have an exception: The three mirrors are of the same type (Fig. 52) because $n = 2$, that is, they each contain two twofold axes. Dihedral angles: $90°$.

Chapter 1   Groups of Isometries

Table 19

Symbol	n^a	Id.	Rotat.	Reflections	Rot.-Inv.	Centr. Inv.	Total
$D_n \times I$	Even	1	$2n - 1$	$n + 1$	$n - 2$	1	$4n$
	Odd	1	$2n - 1$	n	$n - 1$	1	$4n$
$D_{2n}D_n$	Even	1	$2n - 1$	n	n	—	$4n$
	Odd	1	$2n - 1$	$n + 1$	$n - 1$	—	$4n$

^{a} $n \geq 2$ (for $n = 1$: see Example 1).

Table 20

Group	n	Mirrors	2-Fold	n-Fold
$D_n \times I$	Even	1	n	—
		n	1	1
	Odd	n	—	1
$D_{2n}D_n$	Even	n	—	1
	Odd	1	n	—
		n	1	1

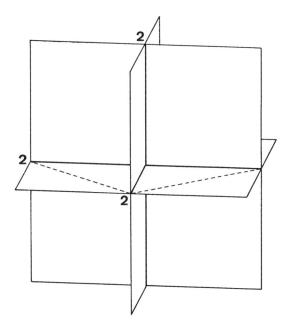

Figure 51. $D_2 \times I$

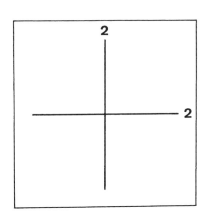

Figure 52. Mirror of $D_2 \times I$, perpendicular to a twofold axis ($3\times$)

b. D_4D_2
   Order 8: E
   　　　　3 rotations
   　　　　3 reflections
   　　　　2 rotatory inversions　(Fig. 53)
   The two mirrors are of the same type with respect to their contents of only one twofold axis. The two remaining axes are bisectors of both planes. Dihedral angles: 90°.
3. Extensions of D_3
   a. $D_3 \times I$
      Order 12: E
      　　　　　I
      　　　　　5 rotations
      　　　　　3 reflections
      　　　　　2 rotatory inversions　(Fig. 54)

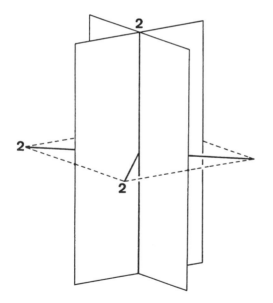

Figure 53. D_4D_2

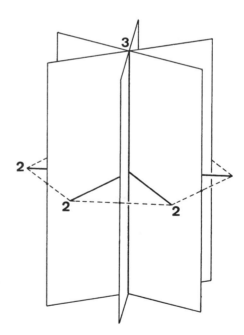

Figure 54. $D_3 \times I$

The three mirrors are of the same type with respect to their contents of the threefold axis only, whereas the three twofold axes are bisectors. Dihedral angles: 60°.

b. D_6D_3

Order 12: E

5 rotations
4 reflections
2 rotatory inversions   (Fig. 55)

Number of mirrors: 3 (Fig. 56a) + 1 (Fig. 56b); dihedral angles: 60°, 90°.

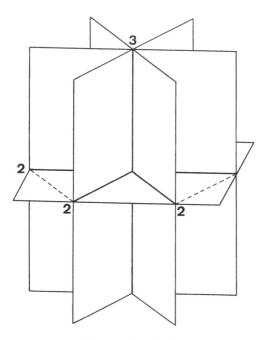

Figure 55. D_6D_3

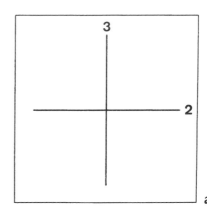

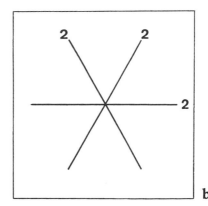

Figure 56. Mirrors of D_6D_3: (a) (3×); (b) perpendicular to the threefold axis

4. Extensions of D_4
   a. $D_4 \times I$
      Order 16: E
                   I
                   7 rotations
                   5 reflections
                   2 rotatory inversions   (Fig. 57)

      Number of mirrors: 4 (Fig. 58a) + 1 (Fig. 58b); dihedral angles: 45°, 90°.
   b. $D_8 D_4$
      Order 16: E
                   7 rotations
                   4 reflections
                   4 rotatory inversions   (Fig. 59)

      The four mirrors contain the fourfold axis only, whereas the four twofold axes are bisectors. Dihedral angles: 45°, 90°.

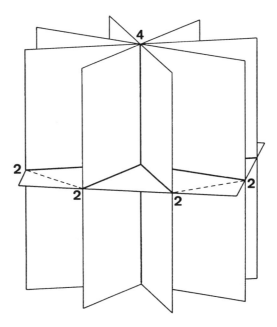

Figure 57. $D_4 \times I$

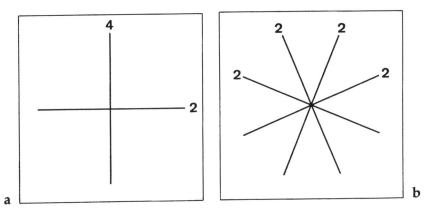

Figure 58. Mirrors of $D_4 \times I$: (a) perpendicular to a twofold axis ($4\times$); (b) perpendicular to the fourfold axis ($1\times$)

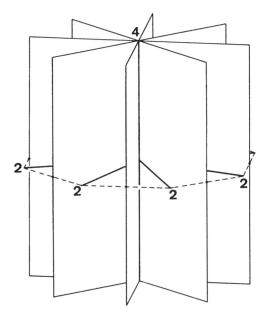

Figure 59. $D_8 D_4$

5. Extensions of D_5
   a. $D_5 \times I$
      Order 20: E
               I
               9 rotations
               5 reflections
               4 rotatory inversions   (Fig. 60)

      The five mirrors contain the fivefold axis only, whereas the five twofold axes are bisectors. Dihedral angles: 36°, 72°.
   b. $D_{10}D_5$
      Order 20: E
               9 rotations
               6 reflections
               4 rotatory inversions   (Fig. 61)

      Number of mirrors: 5 (Fig. 62a) + 1 (Fig. 62b); dihedral angles: 36°, 72°, 90°.

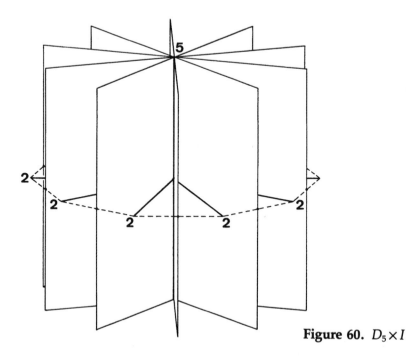

Figure 60. $D_5 \times I$

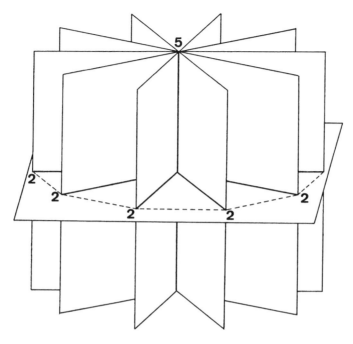

Figure 61. $D_{10}D_5$

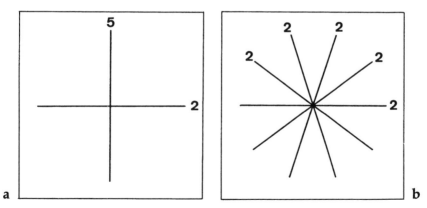

Figure 62. Mirrors of $D_{10}D_5$: (a) (5×); (b) perpendicular to the fivefold axis (1×)

5.3. Tetrahedral

1. $A_4 \times I$

Order 24: E
I
11 rotations
3 reflections
8 rotatory inversions   (Fig. 63)

The three mirrors (Fig. 64) are all alike and contain two twofold axes. The three reflections occur in a subgroup $D_2 \times I$ and their mirrors have dihedral angles of 90°. The four threefold axes appear in the middle of each quadrant. Figure 65 shows spherical sections.

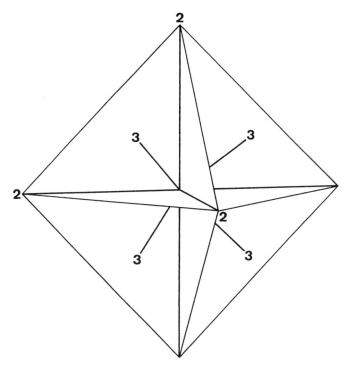

Figure 63. $A_4 \times I$ inside an imaginary octahedron

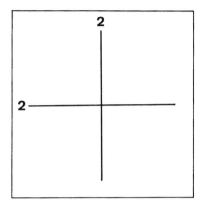

Figure 64. Mirror of $A_4 \times I$, perpendicular to a twofold axis (3×)

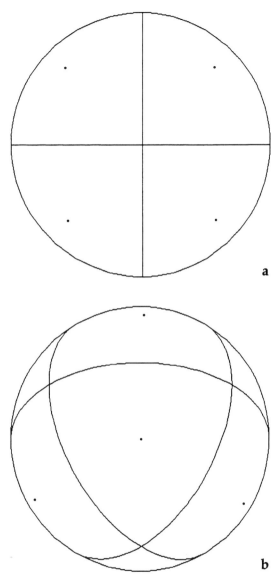

Figure 65. Views of $A_4 \times I$ on a sphere: (a) along a twofold axis; (b) along a threefold axis

2. S_4A_4

Order 24: E
  11 rotations
  6 reflections
  6 rotatory inversions   (Fig. 66)

When subtracting A_4 from S_4, the half-turn in each of the three cyclic subgroups C_4 in S_4 is removed, but the half-turns about the six twofold axes remain, offering six reflections altogether. The six mirrors are all alike (Fig. 67). Dihedral angles: 60°, 90°. Figure 68 shows spherical sections.

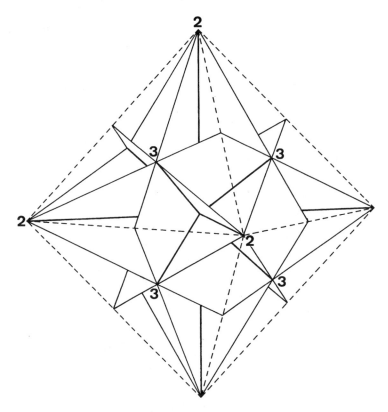

Figure 66. S_4A_4 inside an imaginary octahedron

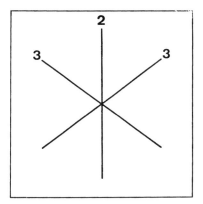

Figure 67. Mirror of S_4A_4 (6×)

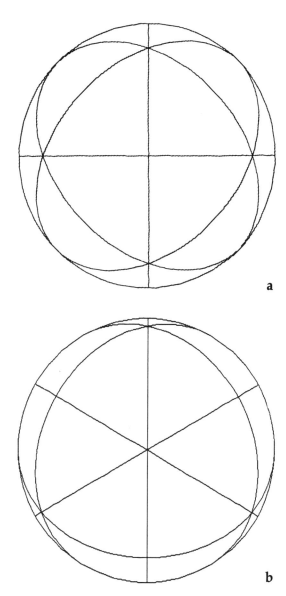

Figure 68. Views of S_4A_4 on a sphere: (a) along a twofold axis; (b) along a threefold axis

5.4. Octahedral

$S_4 \times I$

Order 48: E
I
23 rotations
9 reflections
14 rotatory inversions   (Fig. 69)

The nine mirrors consist of the three mirrors of the subgroup $A_4 \times I$ (Fig. 70a) and the six mirrors of the subgroup $S_4 A_4$ (Fig. 70b). The first three are each perpendicular to a fourfold axis, and the latter six to a twofold axis. Dihedral angles: 45°, 60°, 90°. Figure 70 shows mirrors and Figure 71 spherical sections.

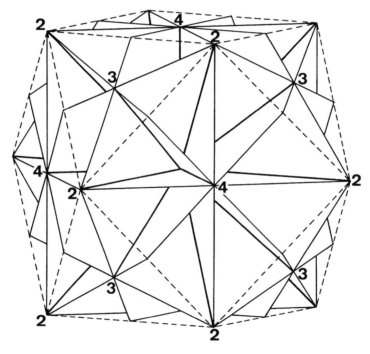

Figure 69. $S_4 \times I$ inside an imaginary cuboctahedron

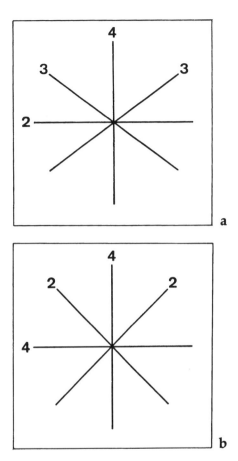

Figure 70. Mirrors of $S_4 \times I$: (a) perpendicular to a twofold axis (6×); (b) perpendicular to a fourfold axis (3×)

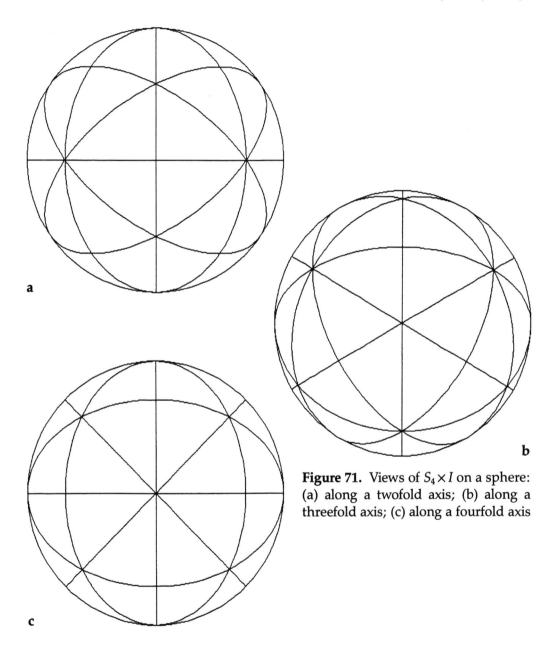

Figure 71. Views of $S_4 \times I$ on a sphere: (a) along a twofold axis; (b) along a threefold axis; (c) along a fourfold axis

5.5. Icosahedral

$A_5 \times I$

Order 120: E
I
59 rotations
15 reflections
44 rotatory inversions   (Fig. 72)

The 15 mirrors are all alike (Fig. 73); dihedral angles: 36°, 60°, 72°, 90°. Figure 74 shows spherical sections.

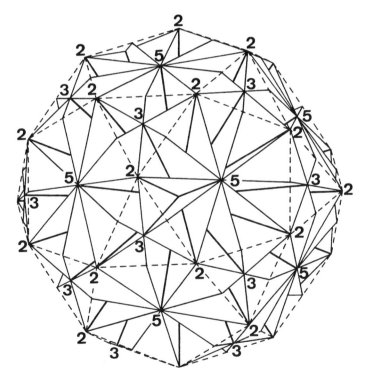

Figure 72. $A_5 \times I$ inside an imaginary icosidodecahedron

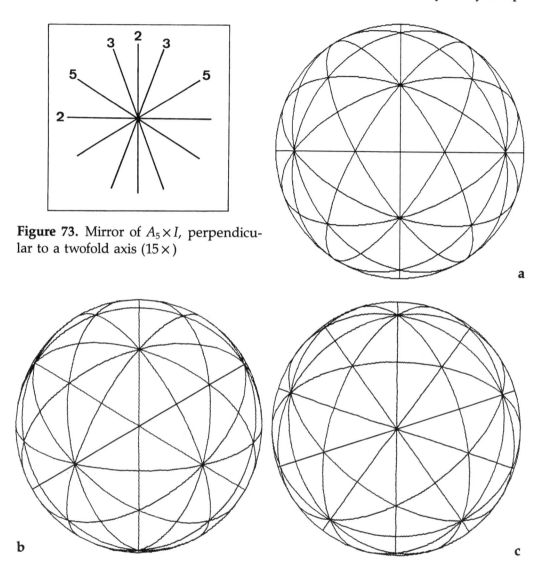

Figure 73. Mirror of $A_5 \times I$, perpendicular to a twofold axis (15×)

Figure 74. Views of $A_5 \times I$ on a sphere: (a) along a twofold axis; (b) along a threefold axis; (c) along a fivefold axis

Chapter 2

Symmetry Action

The gravitational force of the earth remains only an abstract idea as long as there is no object which it is attracting. Similarly, the sun is shining, but when there is no object reflecting the sun-rays, it remains dark and cold out in space.

These comparisons are meant as an introduction to an algebraic idea known as the action of a group on a set, by which the set is being subject to a grouplike structural behavior.

1. The Action of a Group on a Set

In this section, a summary is given of common definitions and group properties which will be further applied.

1.1. Quotient Sets

Let **G** be a group and **H** ⊂ **G** be a subgroup. $C \in$ **G** determines a *right coset* $C \cdot$ **H** and a *left coset* **H** $\cdot C$ of **H**.* Each element of a coset determines the coset itself. **H** is then the right and left coset determined by E, the neutral element in **G**. All cosets are equipotent, i.e., a bijection exists between any two of them. The set of right cosets is denoted by **G**/**H** and the set of the left cosets by **H****G**, which are respectively known as the *right and left quotient set*, both being equipotent. If $C \cdot$ **H** = **H** $\cdot C$ for any $C \in$ **G**, **H** is called a *normal* subgroup in **G**. For example, this occurs for

---

* There is no unanimity concerning the choice of which coset has to be referred to as right or left. Some authors call **H** $\cdot C$ a right coset and $C \cdot$ **H** a left coset.

any incidental subgroup when **G** is commutative. **G**/**H** and **H****G** each consists of a *partition* of **G**: The right or left cosets are *disjunct*, i.e., any mutual intersection is empty and either sum is **G**. If a quotient set is finite, #**G**/**H** is called the *index* of **H** in **G**. Moreover, if **G** is finite, #**G** = g, #**H** = h, and #**G**/**H** = g/h. Some particular normal subgroups follow:

1. **H** = $\{E\}$, the *trivial* subgroup. #**H** = 1 and **G**/**H** = **G**. If **G** is finite, the index is $g/1 = g$.
2. **H** = **G**, the *improper* subgroup. There is only one coset, namely **G**, and the index is 1. If **G** is finite, $h = g$ and the index is $g/g = 1$.
3. **H** is a subgroup of index 2. **H** has now one distinct coset, which is clearly a right as well as a left coset.

1.2. Conjugate Subgroups

An element $A \in$ **G** determines a *conjugate subgroup* $A \cdot$ **H** $\cdot A^{-1}$ of **H** which is identical for all elements of a right coset:

$$(C \cdot \mathbf{H}) \cdot \mathbf{H} \cdot (\mathbf{H}^{-1} \cdot C^{-1}) = C \cdot \mathbf{H} \cdot C^{-1}$$

The set of all conjugate subgroups of **H** is a *class* (*of conjugate subgroups*):

$$\{A \cdot \mathbf{H} \cdot A^{-1} | A \in \mathbf{G}\}$$

This does not necessarily involve a bijection between the right or left quotient set and the class of conjugate subgroups. For example when **H** is normal in **G**, $A \cdot$ **H** $\cdot A^{-1} =$ **H** for any $A \in$ **G**. Then, the class of **H** contains only **H**: Such a subgroup is known as being *self-conjugate*.

1.3. "Activating" a Set

If V is a set, the *product set* of **G** and V is

$$\mathbf{G} \times V = \{(A, p) | A \in \mathbf{G}, p \in V\}$$

and an *action* of the group **G** on the set V is defined as a mapping f

$$f: \mathbf{G} \times V \longrightarrow V$$
$$(A, p) \longmapsto A * p$$

such that $E * p = p$ and $A * (B * p) = (A \cdot B) * p$. This mapping turns V into a *homogeneous space*, which can be seen as a set that reflects something of the structure of the group. However, V has absolutely not become a group: There is no neutral element, nor do there occur inverse elements. As a special case, let V be the set of elements in **G**: **G** is then a homogeneous space also.

1.4. G-Sets

The mappings:

$$q: \mathbf{G} \longrightarrow \mathbf{G}/\mathbf{H} \qquad q': \mathbf{G} \longrightarrow \mathbf{H} \backslash \mathbf{G}$$
$$C \longmapsto C \cdot \mathbf{H} \qquad C \longmapsto \mathbf{H} \cdot C$$

are the *quotient mappings* (q on the right quotient set and q' on the left quotient set) for the subgroup **H**. q determines an action of **G** on **G**/**H**, when $V = \mathbf{G}/\mathbf{H}$, and

$$A * (B \cdot \mathbf{H}) = (A \cdot B) \cdot \mathbf{H}$$

G/**H** is now an example of a homogeneous space, known as a *G-set*. If confusion is to be avoided, G-sets will have to be called right or left, according to the quotient space.

Hence, if **G** is considered as a homogeneous space, the quotient mapping **q** is an epimorphism between two homogeneous spaces which becomes an isomorphism when $H = \{E\}$. **G**/**H** and **G** may then be identified. When **H** is normal in **G**, $\mathbf{G}/\mathbf{H} = \mathbf{H}\backslash\mathbf{G}$. Right and left quotient spaces are now identical and become a *quotient group* for the group multiplication. The neutral element is then **H**.

1.5. Orbits in a Group Action

The *stabilizer* of an element $p \in V$ is the subgroup $\mathbf{H} \subset \mathbf{G}$:

$$\mathbf{H} = \{A \in \mathbf{G} | A * p = p\}$$

Since $E \in \mathbf{H}$, $\mathbf{H} \neq \emptyset$.

The *orbit* of p for **G** is the subset $\mathbf{G} * p \subset V$:

$$\mathbf{G} * p = \{A * p \mid A \in \mathbf{G}\}$$

p shall be said to *describe* the orbit. The elements of the orbit are the *constituents*, among which p is the *descriptive*.

The action group can further be related to the orbit of a descriptive by a surjective mapping r:

$$r: \mathbf{G} \longrightarrow \mathbf{G} * p$$
$$A \longmapsto A * p$$

which shall be called the *realization* of **G** for p. $\mathbf{G} * p$ is then also a homogeneous space.

The realization maps all elements of a right coset of the stabilizer of the descriptive onto one constituent in the orbit:

$$r(\mathbf{H}) = p$$
$$r(C \cdot \mathbf{H}) = (C \cdot \mathbf{H}) * p = C * (\mathbf{H} * p) = C * p$$

Hence, a bijection s exists between the right quotient space of the stabilizer and the orbit of p, which is an isomorphism of homogeneous spaces:

$$s: \mathbf{G}/\mathbf{H} \longrightarrow \mathbf{G} * p$$
$$C \cdot \mathbf{H} \longmapsto C * p$$

or, abbreviated

$$\mathbf{G}/\mathbf{H} \cong \mathbf{G} * p$$

Hence, $\mathbf{G}/\mathbf{H}$ and $\mathbf{G} * p$ may be identified as homogeneous spaces, and p may also be referred to as the descriptive of $\mathbf{G}/\mathbf{H}$.

The following mapping diagram can be made:

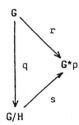

Consequently, if **G** is finite, the number of constituents is the index of the stabilizer.

1.6. Operator Action

A special group action is sent when **G** is a group of operators, V is the set in which the operation of **G** is defined, and $A * p = A(p)$. The orbit of a descriptive is now the homogeneous space of its images under **G**, and the stabilizer is the subgroup of operators that leave the descriptive invariant under **G**. This kind of action shall be called the *operator action* of **G**. For our scope, from here on **G** will be a group of isometries (any subgroup in **I**) and V will be the three-dimensional Euclidean space E^3.

2. Set-Related Definitions and Properties

The operator action of a group **G** of isometries onto E^3 will be discussed further. First, to avoid repetition of similar words, such as "set," "sets of subsets," etc., an adapted terminology particularly suitable for the theoretical aspects of the following subsections needs to be introduced.

2.1. Subset Spaces of E^3 on Higher Levels

If V is a set, the set of all subsets in V shall be denoted by $V^{(1)}$:

$$V^{(1)} = \{W | W \subset V\}$$

$V^{(1)}$ contains also $\emptyset$, V, and all the elements of V as *singletons*:

$$W \in V \Leftrightarrow \{W\} \in V^{(1)}$$

This definition can be applied to $V^{(1)}$. $V^{(2)}$ then denotes the set of all the subsets in $V^{(1)}$:

$$V^{(2)} = \{W | W \subset V^{(1)}\}$$

In steps, a *subset space of level n*, denoted by $V^{(n)}$, is thus obtained whose elements are all the subsets in $V^{(n-1)}$:

$$V^{(n)} = \{W | W \subset V^{(n-1)}\}$$

By extrapolation, V is identified with $V^{(0)}$. $V^{(n)}$ contains $V^{(n-1)}$ and all its elements as singletons. Repeating this for $V^{(n-1)}$, $V^{(n-2)}$ and all its elements are double singletons in $V^{(n)}$ and are denoted between two accolades:

$$W \in V^{(n-2)} \Leftrightarrow \{\{W\}\} \in V^{(n)}$$

Hence, if $0 \leq m \leq n - 1$, $V^{(m)}$ and all its elements are multiple singletons in $V^{(n)}$, and are denoted between $(n - m)$ accolades,

$$W \in V^{(m)} \Leftrightarrow \underbrace{\{\{\{ \cdots }_{n-m} \{W\} \underbrace{\cdots \}\}\}}_{n-m} \in V^{(n)}$$

When $V = E^3$, an element of $V^{(n)}$, being the three-dimensional Euclidean subset space of level n, shall be called a *space-(n) body*. A space-(n) body is then a set of space-$(n - 1)$ bodies, which shall be called its *components*. The sum of these components is again a space-$(n - 1)$ body and is composed of space-$(n - 2)$ bodies. In downward steps, the component sum can be made until level (2), where a space-(1) body is obtained. A component sum appears now on each level (m) $(1 \leq m < n)$, and the set of these $(n - 1)$ bodies, increased with the space-(n) body itself, shall be called the *space-(n) system*, determined by the space-(n) body. The component sum on level $(m + 1)$, offering a space-(m) body, shall then be called the *level-(m) body in the system*, whereas the level-(n) body in the system is the space-(n) body itself. If a level-(m) body is composed of disjunct components, the system shall be called *disjunct* on level (m), and otherwise *conjunct* on this level.

2.2. Isometries on Higher Levels

An isometry is defined in the points of E^3, and, hence, a set in E^3 is transformed into the mapped points of the set. As such, a level-(1) body is transformed into a congruent body. This extension principle can be applied on any level but has yet to be properly defined in steps.

Again, let **I** denote the total group of isometries, and $A \in \mathbf{I}$. The image of a space-(1) body p_1 under A is the congruent body, composed of all the point images:

$$A(p_1) = \{A(P) | P \in p_1\}$$

An operator $A^{(1)}$ is then defined in $V^{(1)}$:

$$A^{(1)}: p_1 \longmapsto A(p_1)$$

The set

$$\mathbf{I}^{(1)} = \{A^{(1)} | A \in \mathbf{I}\}$$

can be subject to a multiplication,

$$A^{(1)} * B^{(1)} = (A \cdot B)^{(1)}$$

turning $\mathbf{I}^{(1)}$ into a group, isomorphic with **I**.

Next, the image of a space-(2) body p_2 under $A^{(1)}$ is similarly defined as the body composed of all the images of its space-(1) bodies p_1:

$$A^{(1)}(p_2) = \{A(p_1) | p_1 \in p_2\}$$

An operator $A^{(2)}$ is then defined in $V^{(2)}$:

$$A^{(2)}: p_2 \longmapsto A^{(1)}(p_2)$$

Similarly, as explained for $\mathbf{I}^{(1)}$, $\mathbf{I}^{(2)}$ is a group, isomorphic with **I**:

$$A^{(2)} * B^{(2)} = (A \cdot B)^{(2)}$$

By induction, the image under $A^{(n-1)}$ of a space-(n) body p_n, composed of space-($n - 1$) bodies p_{n-1}, is defined as

$$A^{(n-1)}(p_n) = \{A^{(n-1)}(p_{n-1}) | p_{n-1} \in p_n\}$$

An operator $A^{(n)}$ is then defined in $V^{(n)}$:

$$A^{(n)}: p_n \longmapsto A^{(n-1)}(p_n)$$

an operator of the group $\mathbf{I}^{(n)}$, isomorphic with **I**.

Therefore, when an isometry A is said to transform a space-(n) body, the associated operator $A^{(n)} \in \mathbf{I}^{(n)}$ is meant. The addition (n) is then deleted, when no confusion occurs: further on, the notation $A(p)$, where p is a space-(n) body, will be referring to $A^{(n)}(p)$. Thus, the operation of an isometry has now been extended to a subset space of any level. As a consequence, the image of a space-(n) system under A is the new system, composed of the n transformed bodies.

3. Descriptive Space-(n) Bodies

The operator action of a group **G** of isometries on E^3 automatically involves an operator action on an E^3 subset space of any level. The *symmetry group* **S** of a space-(n) body $p \in V(n)$ is defined as the stabilizer of p in **I**.

If $\mathbf{S} \in \{E\}$, p shall be called *asymmetrical*. Hence, **S** is also the symmetry group of any level-(m) body in the system of p and may, therefore, be called the symmetry group of the system also.

The stabilizer **H** of p in **G** is then

$$\mathbf{H} = \mathbf{S} \cap \mathbf{G}$$

and may now also be called *the stabilizer of the system*.

$\mathbf{G}(p)$ is a space-($n + 1$) body, isomorphic with the **G**-set **G**/**H**.

4. Positioning in Space

If $\mathbf{G} = \mathbf{I}$, $\mathbf{I}/\mathbf{S} \cong \mathbf{I}(p)$, *the total orbit of p*, whose constituents are all the bodies congruent with p. From here on, a constituent of $\mathbf{I}(p)$ shall be referred to as *a position of p*. Hence, a position may be direct or opposite. Also, the expression "body in a certain position" shall be used.

If $A \in \mathbf{I}$, the symmetry group of $A(p)$ is the group $A \cdot \mathbf{S} \cdot A^{-1}$:

$$A \cdot \mathbf{S} \cdot A^{-1}(A(p)) = A(p)$$

a representative of the class of the conjugate subgroups of **S** in **I**. The stabilizer of $A(p)$ in **G** is

$$A \cdot \mathbf{S} \cdot A^{-1} \cap \mathbf{G}$$

5. Orbit Properties

5.1. Invariance of the Orbit

$\forall A \in \mathbf{G}$:

$$\mathbf{G}(A(p)) = (\mathbf{G} \cdot A)(p) = \mathbf{G}(p)$$
$$A(\mathbf{G}(p)) = (A \cdot \mathbf{G})(p) = \mathbf{G}(p)$$

which means that the orbit of any constituent is identical and that $\mathbf{G}$ is a subgroup of the symmetry group of the space-$(n + 1)$ body $\mathbf{G}(p)$.

5.2. Suborbits

A proper subgroup $\mathbf{G}_s \subset \mathbf{G}$ defines a *subaction* of $\mathbf{G}$, i.e., the action of $\mathbf{G}$, restricted to operators in $\mathbf{G}_s$. Accordingly, $\mathbf{G}_s(p)$ is called the *suborbit* for $\mathbf{G}_s$ in $\mathbf{G}(p)$.

Because $\{E\} \subset \mathbf{G}$ is the trivial subgroup, the suborbit $\{E\}(p) = \{p\}$ is the *trivial suborbit* in $\mathbf{G}(p)$. If $\mathbf{G}_s(p) = \mathbf{G}(p)$, the suborbit is called *improper*. If $\mathbf{H}_s$ denotes the stabilizer of $\mathbf{G}_s(p)$,

$$\mathbf{G}_s(p) \cong \mathbf{G}_s/\mathbf{H}_s$$
$$\mathbf{H}_s = \mathbf{H} \cap \mathbf{G}_s$$

$\mathbf{G}_s(p)$ is now also a space-$(n + 1)$ body. Let p_s denote $\mathbf{G}_s(p)$. $\mathbf{G}(p_s)$ is then a space-$(n + 2)$ body whose components shall be called the *right coorbits* of $\mathbf{G}_s(p)$. Hence, $\mathbf{G}_s(p)$ is one of the right coorbits itself.

Because the realization maps the right cosets of $\mathbf{G}_s$ onto the right coorbits $\mathbf{G}_s(p)$, the mapping

$$t: \quad \mathbf{G}/\mathbf{C}_s \longrightarrow \quad \mathbf{G}_s(p) = p_s$$
$$C \cdot \mathbf{G}_s \longmapsto C \cdot \mathbf{G}_s(p) = C(p_s)$$

relates the right cosets with the right coorbits. Similarly, a mapping t' may relate the left cosets with the left coorbits. If $\mathbf{G}_s$ is normal in $\mathbf{G}$, the right and left cosets are identical, and, hence, the right and left coorbits are identical also, which may then be simply called coorbits.

The mapping t can be either a surjection or a bijection. If the stabilizer of the descriptive p_s is identical to G_s, t is identical to the mapping s (see Section 1.5) for p_s and is then a bijection and vice versa. Hence, if G_1 is a proper subgroup of the stabilizer of p_1, at least two right coorbits are identical, and vice versa.

$G(p_s)$ determines a space-$(n+2)$ system whose level-$(n+1)$ body is identical to $G(p)$. According to whether the system is conjunct or disjunct on level $(n+2)$, the suborbit is called *conjunctive* or *disjunctive*, respectively.

As H and G_s are each proper subgroups in G, three cases are distinguished:

1. $H \subset G_s$. Hence, $H_s = H$ and G/H_s is a refinement of G/G_s. The suborbit is disjunctive, t is a bijection, and G_s is the stabilizer of p_s.
2. $G_s \subset H$ is proper. In this trivial case, the suborbit is clearly conjunctive because t maps the subset $H/G_s \subset G/G_s$ onto H, implying that t is surjective.
3. $G_s \not\subset H$ and $H \not\subset G_s$. According to case 1, the subgroup $\langle H, G_s \rangle \subset H$, generated by H and G_s, is disjunctive.

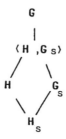

Restricted to $\langle H, G_s \rangle$, the set of right coorbits of $G_s(p)$, sharing one constituent, is equipotent with H/H_s. Because $H_s \subset H$ is proper, there are at least two right cosets in H/H_s, implying that the right coorbits are not disjunct. Hence, $G_s(p)$ is conjunctive, and t may be bijective or surjective.

Moreover, if $G/G_s \leftrightharpoons H/H_s$, $\langle H, G_s \rangle = G$. Hence,

$$G_s(p_s) = G(p_s) = p_s$$

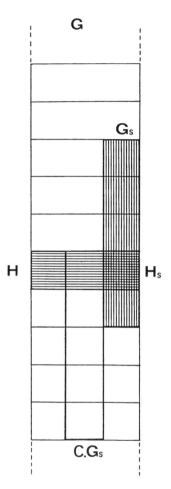

Figure 75. The subgroups H_1, H_2, and G_1 with cosets in G_2

In conclusion: a suborbit $G_s(p)$ is disjunctive if $H \subset G_s$ and conjunctive if $H \subset G_s$. If $G(p)$ contains no suborbit which is proper, nontrivial, and disjunctive, it is called *prime*. H is a *direct* subgroup in G, when G contains no proper subgroup between H and G. Every incidental suborbit is then either trivial and improper or conjunctive, implying that $G(p)$ is prime.

5.3. The Symmetry Group of an Orbit

If **Z** denotes the symmetry group of **G**(p), Diagram 2 can be applied when G_s and **G** are replaced by **G** and **Z**, respectively, and H_s and **H** by **H** and H_Z, respectively.

Now **G**(p) = **Z**(p). Hence, the right coorbits of **G**(p) in **Z**(p) are obviously coincidental, and

$$G/H \leftrightharpoons Z/H_Z$$
$$Z/G \leftrightharpoons H_Z/H$$

6. Descriptive Space-(1) Bodies

In this section, special attention will be dedicated to descriptives in $V^{(1)}$, which are either singletons, containing one point, or sets of more than one point.

6.1. Enantiomorphous Coorbits

If the index of G_s is 2, the two coorbits are either disjunct (Section 5.2, item 1) or identical (Section 5.2, item 3).

Every group **G**, containing opposite isometries has a subgroup of index 2; the subgroup G_d of all the direct isometries. Coorbits for G_d shall be called *enantiomorphous*. The suborbit $G_d(p)$ contains all the constituents under direct isometries, and its coorbit contains all those under opposite isometries. If an opposite image occurs in the latter one, the suborbit is definitively disjunctive. Let $A \in \mathbf{G}$ be opposite:

$$A(p) \text{ opposite} \Leftrightarrow \forall B \in \mathbf{G}_d: \quad B(p) \neq A(p)$$
$$A^{-1} \cdot B(p) \neq p$$
$$B \notin A \cdot \mathbf{S}$$

Hence, $A(p)$ is only opposite if **S** does not contain opposite isometries. Let **G** contain opposite isometries:

1. If p has opposite symmetry, no opposite constituents occur in the orbit.
2. If p lacks opposite symmetry and is positioned according to Section 5.2, item 1, the suborbit contains the direct constituents and the coorbit contains the opposite constituents.

6.2. Decomposition Sequences

Any countable decomposition sequence of **G** can be considered as a sequence of n groups:

$$\mathbf{G}_0 \subset \mathbf{G}_1 \subset \mathbf{G}_2 \cdots \subset \mathbf{G}_{n-2} \subset \mathbf{G}_{n-1}$$

where $\mathbf{G}_0 = \{E\}$ and $\mathbf{G}_{n-1} = \mathbf{G}$. Let p be denoted by p_1 and have a stabilizer **H** in **G** and $\mathbf{H}_1$ in $\mathbf{G}_1$. This sequence associates a space-(n) system constructed from p as follows:

p_1 is the descriptive of $\mathbf{G}_1/\mathbf{H}_1$ and $p_2 = \mathbf{G}_1(p_1)$ is a space-(2) body, with a stabilizer $\mathbf{H}_2$ in $\mathbf{G}_2$. p_2 is then the descriptive of $\mathbf{G}_2/\mathbf{H}_2$ and $p_3 = \mathbf{G}_2(p_2)$ is a space-(3) body with a stabilizer $\mathbf{H}_3$ in $\mathbf{G}_3$. If $m < n - 1$, by induction, p_{m-1} is the descriptive of $\mathbf{G}_{m-1}/\mathbf{H}_{m-1}$ and $p_m = \mathbf{G}_{m-1}(p_{m-1})$ is a space-(m) body, with a stabilizer $\mathbf{H}_m$ in $\mathbf{G}_m$. If $m = n - 1$, p_n is the final space-(n) body in this sequence and determines a space-(n) system. This system has the property that each level-(m) body in it is identical with $\mathbf{G}/\mathbf{H}_{m-1} \cong \mathbf{G}(p_{m-1})$ ($m \neq 1$).

A decomposition sequence, where $\mathbf{H} = \mathbf{G}_0$ or, if **H** is nontrivial, $\mathbf{H} = \mathbf{G}_1$, involves a completely disjunct system. An interesting complexity of multiple realizations is then seen through the orbit $\mathbf{G}(p)$, when the various space-(2) descriptives are the suborbits $\mathbf{G}_m(p)$. The level-(2) body of the space-(m) system, determined by p_m, is the suborbit $\mathbf{G}_m(p) \subset \mathbf{G}(p) \cdot \mathbf{G}_m/\mathbf{H} \cong \mathbf{G}_m(p)$ and $\mathbf{G}_m(p)$ is the descriptive of $\mathbf{G}/\mathbf{G}_m$ because its stabilizer in **G** is $\mathbf{G}_m$. Such a sequence shall be called a *disjunctive decomposition sequence of* $\mathbf{G}(p)$.

Chapter 3

Orbit Systems

In this chapter, some geometrical application possibilities of the realization theory in the previous chapter will be explained. It will be shown how different sorts of polyhedral symmetric shapes are actually associated with realizations of groups of isometries. According to the given definitions, a group realization implies a homogenous space isomorphism between a right **G**-set and the orbit of the descriptive. The right **G**-set is determined by the stabilizer of the positioned descriptive. Examples of descriptives will be restricted to space-(1) bodies, whereas in Chapter 6, some attention will be also dedicated to higher-level descriptives. Two distinct categories occur on level (1): point singletons and sets of more points.

1. Descriptive Singletons

The descriptive $p = \{P\}$, where $P \in E^3$, is a space-(1) body composed of one point only. The set of points in the singletons of the orbit $\mathbf{G}(p)$ is then generally incoherent.

1.1. Examples

1. Cyclic Groups of Translations

Let T denote a translation in E^3. The group $\langle T \rangle$, generated by T, is a discrete cyclic subgroup in **I**, which has no points of invariancy. Hence, the stabilizer of any p is trivial. The points in the singletons of $\mathbf{G}(p)$ are distributed as an infinite sequence on a line (Fig. 76), where two successive points are equidistant.

Figure 76. $\{P\}$ describes a discrete group, generated by a translation

2. Groups of Rotations Having One Point of Invariancy

Three position types for P are to be distinguished: (a) in the center of invariancy 0, (b) on an axis, but distinct from 0, and (c) outside the axes.

a. $P \equiv 0$. $\mathbf{H} = \mathbf{G}$ and $\mathbf{G}(p)$ contains p as a single constituent.
b. $P \not\equiv 0$ and P lies on an axis of $\mathbf{G}$. The stabilizer $\mathbf{H}$ is now the cyclic subgroup of rotations, determined by the axis, and p is a descriptive of $\mathbf{G}/\mathbf{H}$; for example, let $\mathbf{G} = A_5$ and P be chosen on a fivefold axis of A_5. Then, $\mathbf{H} = C_5$, $g = \#\mathbf{G} = 60$, and $h = \#\mathbf{H} = 12$. The points in the $g/h = 12$ singleton constituents of $\mathbf{G}(p)$ are distributed as six opposite pairs, lying respectively on each of the six fivefold axes of A_5. This can be easily visualized. There is a twofold rotation $R \in A_5$, whose axis is perpendicular to the fivefold axis of $\mathbf{H}$ (see Table 16). The isomorphism s (see Chapter 2, Section 1.5) maps the right coset $R \cdot \mathbf{H}$ onto the singleton constituent $R(p)$, which contains the opposite point of P along the axis of $\mathbf{H}$, with respect to 0. Thus, each fivefold axis associates two cosets and one conjugate subgroup of $\mathbf{H}$.

As a counterexample, let $\mathbf{G} = A_4$, and P belong to a threefold axis. Then, $\mathbf{H} = C_3$, $g = 12$, and $h = 3$. The points in the $g/h = 4$ singleton constituents of $\mathbf{G}(p)$ are now each lying on one of the four threefold axes, but on one side of 0 only. Of A_4, there is no twofold axis perpendicular to any threefold axis. Each threefold axis associates one coset and one conjugate subgroup of $\mathbf{H}$.

c. P is outside the axes. $\mathbf{G}(p) \cong \mathbf{G}$ (as homogeneous spaces) and p is a descriptive of $\mathbf{G}$.

1.2. Orbit Models

Models of orbits may either be drawn as two-dimensional projections or may be physically constructed in three dimensions. A descriptive singleton, however, is not a good choice for either kind of model; for example, in a singleton description of S_4, the orbit is composed of 24 singleton constituents. A drawing would represent a flat projection, being just an unsurveyable planar arrangement of the 24 points in the singletons, whereas a three-dimensional model would have to be constructed from 24 objects, actually representing singletons, whose points are floating loosely in space. Because points have actually no dimension at all, these objects would have to be cubes, balls, or any other sort of solid shape.

2. Descriptive Sets of More Points

If a descriptive is distinct form a singleton and is an incoherent set of points, the orbit is a space-(2) body, whose level-(1) body is again an incoherent set of points. In practical examples, incoherent constituents in an incoherent orbit would become hard to recognize. Therefore, from here on, a descriptive of this kind will be considered to be coherent.

The space-(2) system, determined by $\mathbf{G}(p)$, is composed of two bodies: the orbit $\mathbf{G}(p)$, as the level-(2) body, and the sum of its constituents, as the level-(1) body. Each of these has a distinct aspect in connection with the orbit.

2.1. Fundamental Descriptives

Let $p, q \in V^{(1)}$, and $q \subset p$ a proper, nontrivial subset. Then, the constituent sum of $\mathbf{G}(q)$ is a subset of the constituent sum of $\mathbf{G}(p)$, but not necessarily proper:

$$\bigcup \{A(q) | A \in \mathbf{G}\} \subset \bigcup \{A(p) | A \in \mathbf{G}\}$$

However, if

(a) the subset *is* improper and
(b) for each proper subset $q' \subset q$,

$$\bigcup \{A(q') | A \in \mathbf{G}\} \subset \bigcup \{A(p) | A \in \mathbf{G}\}$$

is a proper subset,

q shall be called a *fundamental descriptive* of $\mathbf{G}(p)$. A fundamental descriptive is obviously positioned such that its stabilizer in $\mathbf{G}$ is trivial.

Any space-(1) body can be considered as the constituent sum of its own orbit under any subgroup of its symmetry group and, therefore, a fundamental descriptive as a subset of the body exists. As an example, a sphere is invariant under any isometry that leaves its center invariant and has a discrete symmetry group $\mathbf{S}$, in which $\mathbf{G} = A_5 \times I$ is a proper subgroup. A Möbius triangle p (Fig. 77) is a fundamental descriptive of $\mathbf{G}(p)$,

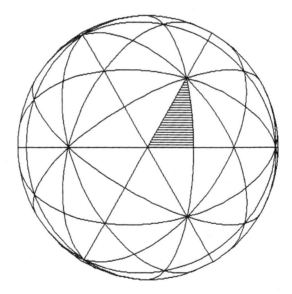

Figure 77. Orbit of a Möbius triangle for $A_5 \times I$, with a trivial stabilizer

whose constituent sum is the sphere. However, a Schwarz triangle—a triangular sum of bounding Möbius triangles—is certainly not. The expression "fundamental descriptive" stands for an extension of the definition of a *fundamental region* [3, 5]. The covering of the sphere with 60 Möbius triangles is also known as an example of a spherical *tesselation*.

2.2. The Orbit System on Level (1)

The sum of the constituents of $G(p)$ is a level-(1) body, which is a *symmetry solid* if it is coherent. As such, a well-applicable construction principle for symmetry solids is provided.

1. Revolution Body

If **G** is a discrete group of rotations, the constituent sum of $G(p)$ is a *revolution body*. A descriptive may clearly be a solid, a curve, or even a planar subset, like a circle or a line segment.

As an example, let the descriptive be a sphere p, positioned such that its center is outside the single axis of **G**. The orbit is an infinite composition of spheres, whose centers are distributed about a circle, perpendicular to the axis, and which is determined by the center of the descriptive. The constituent sum is the three-dimensional area bounded by a *torus*. The symmetry group of the torus contains all the reflections, whose mirrors meet in the axis, plus the one reflection, whose mirror contains the circle of the constituents' centers.

The sphere is clearly not a fundamental descriptive of $G(p)$. A mirror through the axis intersects the torus in two congruent circular areas. If these are disjunct, each of these would be a fundamental descriptive of $G(p)$. Such a descriptive circle involves a disjunct space-(2) system of the orbit. However, this is not a general property; for example, compare with the previous example of the description of $A_5 \times I$ by a Möbius triangle, whose boundary constituents have a common edge. The latter system is conjunct.

2. *Polygon*

Let A be a rotation of finite order a and polygonal value m (10).*

If P is outside the axis of A, let a denote the line segment $[P, A(P)]$ and let $\mathbf{G} = C_s$, the finite cyclic group of rotations generated by A. The constituent sum of $\mathbf{G}(a)$ is a *regular polygon* $\{m\}$ of density s/m. As such, a regular polygon is a broken line segment, whereas the planar area, bounded by it, is the (*polygonal*) *face*. Consequently, a regular polygon is an example of a *finite revolution body*. Yet a is not a fundamental descriptive of $\mathbf{G}(a)$ because boundary constituents share a point. An example of a fundamental descriptive would be $[P, A(P)[$ when $d = 1$.

3. *Dipolygonid–Tripolygonid–Triadic Polyhedron*

Let A and B each be a rotation as in Example 2 whose axes intersect in one point, 0, only. Let ab denote the broken line

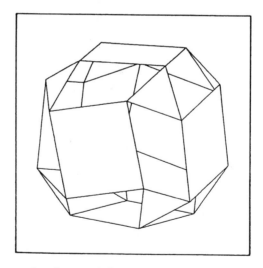

Figure 78. A dipolygonid for S_4, composed of eight triangles and six squares

* $m = s/d$, where d is the *density* of the polygon [3].

Chapter 3 Orbit Systems

segment in P whose edges are $a = [P, A(P)]$ and $b = [P, B(P)]$, and let $\mathbf{G} = \langle A, B \rangle$, the group of rotations generated by A and B which has a single point of invariancy 0. The constituent sum of $\mathbf{G}(ab)$ is an expandable *dipolygonid* representing a generalization of the "jitterbug" [10] structure. This constituent sum is composed of two subsets of each one kind of regular polygons, determined by the polygonal value of A and B, respectively. $\mathbf{G}$ may be finite (Fig. 78) or discrete.

Table 21 is listing the pairs (A, B) generating a finite group $\langle A, B \rangle$. Hence, from Table 21, a classification of the finite dipolygonids follows:

◊ 2 infinite families for the dihedral groups D_n
◊ 2 for A_4
◊ 4 for S_4
◊ 14 for A_5

Next, $B^{-1} \cdot A$ similarly defines a line segment $c = [P, B^{-1} \cdot A]$, which is added to the broken line segment ab, thus forming a "tripod" abc in P. The constituent sum of $\mathbf{G}(abc)$ is then a *tripolygonid*, so-called because it is associated with a triad of axes of $\mathbf{G}$. It is composed of three subsets of each one kind of regular

Table 21. Pairs of rotations with coplanar axes generating a finite group. Numbers represent the order of A and B.

A	B	Angle (θ)	$\langle A, B \rangle$	A	B	Angle (θ)	$\langle A, B \rangle$
2	2	$k/n \cdot 180°$ [a]	D_n	2	n	$90°$	D_n
2	3	$20°54'18''.56$	A_5	3	3	$41°48'37''.12$	A_5
		$35°15'51''.80$	S_4			$70°31'43''.60$	A_4
		$54°44'08''.20$	A_4	3	4	$54°44'08''.20$	S_4
		$69°05'41''.44$	A_5	3	5	$37°22'38''.53$	A_5
2	4	$45°$	S_4			$79°11'15''.65$	A_5
2	5	$31°43'02''.91$	A_5	4	4	$90°$	S_4
		$58°16'57''.09$	A_5	5	5	$63°26'05''.82$	A_5

[a] When $n = 2$, $k = 1$. When $n \geq 3$, $1 \leq k < n/2$; k and n are coprimes.

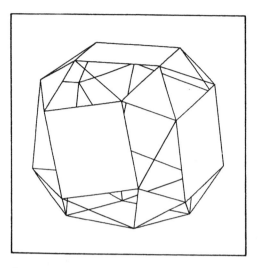

Figure 79. The tripolygonid in triad $S_4(2,3,4)$, derived from the dipolygonid in Fig. 77

polygons, of which the third kind is determined by the polygonal value of $B^{-1} \cdot A$.

In the tripolygonid, *triadic triangles* occur, each of which is bounded by regular polygons of each kind. The set of triadic triangles is equipotent with **G**. If all the faces are added to the tripolygonid's polygons, a *triadic polyhedron* is obtained, which is a distorted, but nevertheless symmetrical version of one of the different types of convex or nonconvex *uniform snub polyhedra* when **G** is finite.

Hence, an alternative but visually interesting way to generate the uniform snub polyhedra is to let a dipolygonid, where $|a| = |b|$ expand and contract until a number of positions is found where $|c| = |a|$; for example, the triadic polyhedron, obtained from Fig. 79, is a *snub cube*.

4. Uniform Polyhedron

In a uniform polyhedron, all faces are regular polygons of equal edge length, and all vertices are alike. The listing of these was prepared by means of the Wythoff–Coxeter construction in the different polyhedral kaleidoscopes determined by Schwarz triangles [5]. Some well-defined positions of points in such a kaleidoscope are given an associated constructive fundamental part, which is repeatedly reflected by the kaleidoscope into a uniform polyhedron—that is, all but one. Each such uniform polyhedron is to be considered as the constituent sum of the orbit of a fundamental descriptive of a finite group of isometries. Among these, the groups of rotations uniquely act for the snub polyhedra.

2.3. The Orbit System on Level (2)

Although the interest of the orbit system on level-(1) was found in the symmetry aspects of the constituent sum under the action group, preferably with a trivial stabilizer of the descriptive, the importance of the orbit itself lies in its isomorphism with a right **G**-set, preferably determined by a *non*trivial stabilizer. As an example, S_4 is described by a sphere which is positioned such that the arrangement of 24 spheres is disjunct (Fig. 80).

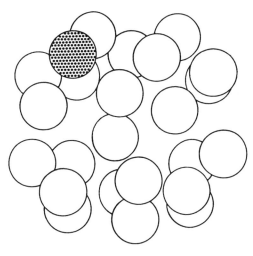

Figure 80. Orbit of a sphere for S_4, with a trivial stabilizer

Because each subgroup of S_4 is also a subgroup of **S**, the sphere can be positioned such that **H** may coincide with any such subgroup. Then, the arrangement of 24 spheres would be reduced to one of 12, 8, 6, 4, 3, 2, or 1 spheres, according to the right **G**-set which the sphere is describing. The latter arrangement—only one sphere—occurs when its center coincides with the point of invariancy of S_4, i.e., when $\mathbf{H} = S_4$. However, should a less symmetrical body be chosen as a descriptive, the centered body would generally describe an orbit of more than one constituent.

3. Compounds of Centered Bodies

If the symmetry group **S** of a space-(1) body, distinct from a singleton, has only one point 0 of invariancy, 0 shall be called the *center* of the body. If the body is positioned such that its center coincides with a point of invariancy of a nontrivial action group of isometries **G**, it shall be called *centered* with respect to the action group. The expression "centered" has thus to be understood in a wider sense. Among the finite groups of isometries, D_1C_1 has a plane of invariancy, C_n and D_nC_n a line of invariancy, and all other groups have a single point of invariancy. All constituents of the orbit are then centered also, and the orbit is called a *compound* of the identical centered bodies.

If $I \in \mathbf{S}$ but $I \notin \mathbf{G}$, $\mathbf{G} \times I$ is a group* in which **G** is a normal subgroup of index 2. Then, the stabilizer in $\mathbf{G} \times I$ is partly outside **G**, which implies that both coorbits are coincidental (see Chapter 2, Section 6.1). Hence, the compound is also invariant under $\mathbf{G} \times I$. As a consequence, when the descriptive has central inverse symmetry, the symmetry group of a compound contains I, and can therefore never be a mixed group.

---

* Actually $\mathbf{G} \cup \mathbf{G} \cdot I$. The notation is used in accordance with the Coxeter notation $\mathbf{G} \times I$ for finite groups [3].

Polyhedral Compounds

In the just recently (1954/1975, see Historical Appendix) established list of uniform polyhedra, each polyhedron has a center. Among these, the following occur without central inverse symmetry and are listed according to their symmetry group:

- ◊ 2 for $S_4 A_4$: the tetrahedron and the truncated tetrahedron
- ◊ 1 for S_4: the snub cube
- ◊ 8 for A_5: the snub dodecahedron and seven nonconvex snub polyhedra

For the dihedral groups ($n \geq 3$), part of the prisms and the antiprisms have no central inverse symmetry, depending on n. Compounds of identical polyhedra are orbits of centered polyhedra under finite groups of isometries and are isomorphic with right **G**-sets. Hence, these compounds can be systematically listed. All that is needed is to compare the subgroups of the symmetry group of the polyhedron with those of the action group and to place the polyhedron in all possible positions, where the stabilizer is a subgroup of the symmetry group. Because of the attention that some cube compounds have had in the past, more than any other compound type, the cube deserves the privilege of being the first descriptive. In Part II, the analysis and listing of the cube compounds will be carried out.

PART II

Compounds of Cubes

Chapter 4

Classification of the Finite Compounds of Cubes

A compound of cubes has been defined as the orbit of a centered cube for an action group **G** of isometries. As such, it is the expression of an isomorphic right **G**-set **G/H**, where **H** is the stabilizer of the descriptive position of the cube. In this part, such orbits will be constructed and classified for the finite groups of isometries, and discussed afterward. As has been explained, however, these groups may be restricted to those, containing I, which will be further referred to as *I-groups* (and *I-subgroups*).

Since a centered cube **c** is invariant under I, any rotatory inversion $A \cdot I$ maps the cube onto $A \cdot I(\mathbf{c}) = A(\mathbf{c})$. Hence, a cube compound is identical to the suborbit for the subgroup of rotations.

Before being fully able to perform a detailed classification of the cube compounds, specific applications of the theoretical treatment in the previous chapter have to proceed. Some questions that require preliminary attention follow:

◊ What is the set of positions with an identical stabilizer?
◊ Does any distinct position provide a different compound?
◊ What is the actual symmetry group of the compound?

1. The Orientation of a Cube

Let the position of a cube in E^3 be denoted by p, and its symmetry group by **S**, a representative of the class $S_4 \times I$. Any *I*-subgroup in $S_4 \times I$ determines also a class of conjugate I-subgroups. Let **F** denote such a representative. When p is invariant under **F**, the position shall be called *oriented* for **F**.

The set A of oriented positions of p for **F** shall accordingly be called the *orientation* of the cube for **F**. An isometry $C \in \mathbf{I}$ determines a representative $C \cdot \mathbf{F} \cdot C^{-1}$ of the class of **F**, for which the orientation of the cube is $C(A)$, the transformed orientation under C. As such, an orientation is also positioned in E^3. The kind of orientation for each I-subgroup class of $S_4 \times I$ must first be determined.

The nine I-subgroup classes of $S_4 \times I$ are represented by $S_4 \times I$, $A_4 \times I$, $D_4 \times I$, $D_3 \times I$, $D_2 \times I$, $C_4 \times I$, $C_3 \times I$, $C_2 \times I$, $E \times I$ ($= \{E, I\}$ or $C_1 \times I$). The orientation of the cube for each of these nine I-groups is distinguished by a specific kind of coaxiality with **S**. The coincidence of an n-fold axis of **S** and an m-fold axis of **F** is denoted by

$$n \mapsto m$$

Each oriented position is centered for the group; the information will be further deleted. The positions in an orientation are said to have a certain *freedom*, which is dependent on the I-group, as further specified.

1.1. Finite Freedom

For dihedral or higher I-groups, the orientation of the cube is a finite set.

1. $S_4 \times I$

Coaxiality: $2 \mapsto 2 \; (6\times)$
$3 \mapsto 3 \; (4\times)$
$4 \mapsto 4 \; (3\times)$
Orientation: 1 position.

Chapter 4 Classification of the Finite Compounds of Cubes

2. $A_4 \times I$

   Coaxiality: $3 \mapsto 3\ (4\times)$
   $\phantom{\text{Coaxiality: }}4 \mapsto 2\ (3\times)$
   Orientation: 1 position.

3. $D_4 \times I$

   Coaxiality: $2 \mapsto 2\ (2\times)$
   $\phantom{\text{Coaxiality: }}4 \mapsto 2\ (2\times)$
   $\phantom{\text{Coaxiality: }}4 \mapsto 4$
   Orientation: 2 positions, being rotated images of one another through 45° about the shared fourfold axis (Fig. 81).

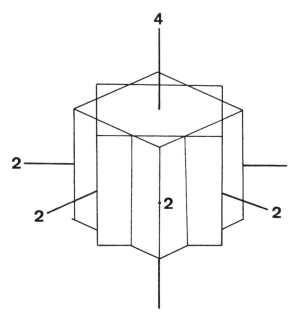

Figure 81. The two cubes in the orientation for $D_4 \times I$

4. $D_3 \times I$

Coaxiality: $2 \mapsto 2\ (3\times)$
$3 \mapsto 3$

Orientation: 2 positions, which are rotated images of one another through 70°31'43".60 about any of the shared twofold axes, or through 60° about the shared threefold axis (Fig. 82).

5. $D_2 \times I$

Coaxiality: (A) $2 \mapsto 2\ (2\times)$
$4 \mapsto 2$
(B) $4 \mapsto 4\ (3\times)$

Orientation: 4 positions, of which 3 are of type (A) and 1 of type (B). Each of the three twofold axes of $D_2 \times I$ is
◊ shared with a pair of cubes, which are rotated positions of one another through 90° about such an axis
◊ a shared fourfold axis of the remaining pair of cubes, about which the cubes are rotated positions of one another through 45° (Fig. 83).

1.2. Rotational Freedom

The orientation of a cube for a cyclic *I*-group is the discrete set of distinct centered positions sharing the group's axis. If any such position p is taken as a reference position, this set is obtained as the rotated positions of p about the axis through the angles $\mu \in [0°, \mu_1[$. This half-open interval shall be applied to determine the orientation.

1. $C_4 \times I$

Coaxiality: $4 \mapsto 4$
Orientation: $[0°, 90°[$

Chapter 4 Classification of the Finite Compounds of Cubes

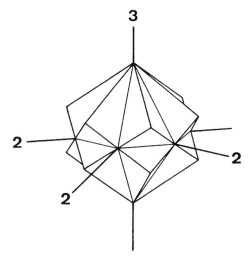

Figure 82. The two cubes in the orientation for $D_3 \times I$

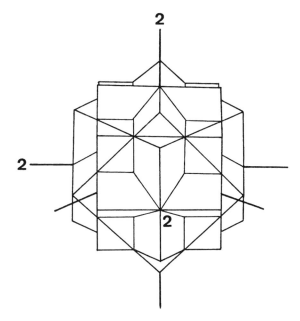

Figure 83. The four cubes in the orientation for $D_2 \times I$

2. $C_3 \times I$

　　Coaxiality: $3 \mapsto 3$
　　Orientation: $[0°, 120°[$

3. $C_2 \times I$

　　Coaxiality: (A) $2 \mapsto 2$
　　　　　　　　(B) $4 \mapsto 2$
　　Orientation: (A) $[0°, 180°[$
　　　　　　　　(B) $[0°, 90°[$

1.3. Central Freedom

Oriented for $E \times I$, a cube is merely centered.

2. Orientative Compounds

2.1. Orientative Stability

Let **F** be a representative of an *I*-subgroup class of $S_4 \times I$, in an action group **G**. The orientation of the cube for **F** is *A*, and a position $p \in A$ determines a stabilizer $\mathbf{H}_p \subset \mathbf{G}$, where $\mathbf{F} \subset \mathbf{H}_p$ is a subgroup in **G**. The compound v, described by p for **G**, is then isomorphic with the right **G**-set $\mathbf{G}/\mathbf{H}_p$.

Therefore, the notation

$$(\mathbf{G}/\mathbf{F})$$

is applied to indicate the *orientative stability* for **F** of a compound for **G**, which refers to a minimal stabilizer **F** in **G** of the descriptive. Such a compound shall be called *orientative* for **F** in **G** and accordingly denoted by

$$(m\,|\,\mathbf{G}/\mathbf{F})$$

which indicates the orientative stability for **F** in **G** and the maximal number m of constituents. These compounds are the first to be classified.

A compound $(m\,|\,\mathbf{G}/\mathbf{F})$ has an *upgraded* orientative stability, which is:

⋄ A *superior orientative stability* $(\mathbf{G}'/\mathbf{F}')$ if at least one diamond of higher symmetry occurs (see Chapter 2, Section 5.3). Among these, $\mathbf{G}'$ is the highest *I*-group, in which the compound is orientative for an *I*-subgroup $\mathbf{F}'$. The orientative compound is then identical to $(m\,|\,\mathbf{G}'/\mathbf{F}')$.
⋄ A *higher orientative stability* $(\mathbf{G}/\mathbf{F}_1)$ if it is also orientative for at least one higher *I*-subgroup in $\mathbf{G}$. Among these, $\mathbf{F}_1$ is the highest *I*-subgroup. The compound is then identical to $(m_1\,|\,\mathbf{G}/\mathbf{F}_1)$, where m_1 is a proper factor of m.
⋄ A *higher superior orientative stability* $(\mathbf{G}'/\mathbf{F}'_1)$ if it has an orientative superior and a higher orientative stability simultaneously. The compound is then identical to $(m_1\,|\,\mathbf{G}'/\mathbf{F}'_1)$. Such cases, however, do not occur.

The *maximal* orientative stability $(\mathbf{G}_m/\mathbf{F}_m)$ is an upgraded orientative stability for the highest *I*-subgroup $\mathbf{F}$ in the highest *I*-group $\mathbf{G}$.

2.2. Versatility

Let the maximal orientative stability of a compound be $(\mathbf{G}/\mathbf{F})$. The set V of distinct compound versions of $(m\,|\,\mathbf{G}/\mathbf{F})$ shall be called the *versatility* of $(m\,|\,\mathbf{G}/\mathbf{F})$. Here, the expression "distinct" is to be understood literally: The compound versions are mutually noncoincidental, but at least one subset of congruent versions may be contained. Therefore, a *fundamental versatility* of $(m\,|\,\mathbf{G}/\mathbf{F})$ is a subset $V_1 \subset V$ of mutually noncongruent versions.

2.3. Stability

Each $v \in V_1$ has a distinct descriptive $p \in A$, with a stabilizer $\mathbf{H}_p \subset \mathbf{F}$. If at least one diamond of higher symmetry exists for p, $\mathbf{G}/\mathbf{H}_p \cong \mathbf{G}'/\mathbf{H}'_p$, and the stabilizer of p is $\mathbf{H}'_p$ in $\mathbf{G}'_p$. Such a right $\mathbf{G}$-set for the highest *I*-group $\mathbf{G}'$ shall be called the *stability* of version v, or of position p. Consequently, if no diamond of higher symmetry occurs, the stability is G/H_p, which in that case is the highest such G-set. With respect to the orientative stability, v has one of the following:

- ◊ A *basic stability* G/F when $F = H_p$.
- ◊ A *superior stability* G'/F' when $F = H_p$ and at least one diamond of higher symmetry occurs. G' is the highest such I-group.
- ◊ A *higher stability* G/H_p when $F \subset H_p$ is proper and no diamond of higher symmetry occurs.
- ◊ A *higher superior stability* G'/H'_p when $F \subset H_p$ is proper and at least one diamond of higher symmetry occurs. G' is the highest such I-group. (Hence, $F' \subset H'_p$ is proper.)

The three latter examples of stability are *upgraded*.

2.4. *I*-Subgroup Diagrams

For each kind of finite I-group of isometries, diagram 1 illustrates the set of I-subgroups. In the left column, the order of I-subgroups is classified by increasing numbers, and the index in a next higher subgroup is indicated. By multiplication of the indices, the index in each higher subgroup is then found.

In the groups $D_{2n} \times I$ ($n \geq 2$) and $S_4 \times I$, two different kinds of subgroups $C_2 \times I$ occur, which shall be distinguished

    a. about a twofold axis: $C_2 \times I$.
    b. about a $2n$-fold axis: $D_1 \times I$.

3. Classification of the Orientative Compounds

The compound versions in the versatility are said to have central, rotational, or rigid freedom when the orientation has central, rotational, or finite freedom, respectively.

The principle of classification is based on the orientative stabilities. The (fundamental) versatility is being indicated by the descriptive positions. The versatility itself is only mentioned when it is distinct from fundamental versatility.

3.1. Rigid Freedom

1. $S_4 \times I$

$(1 | S_4 \times I / S_4 \times I)$
Fundamental versatility: 1 position (see Section 1.1.1)
Stability: basic

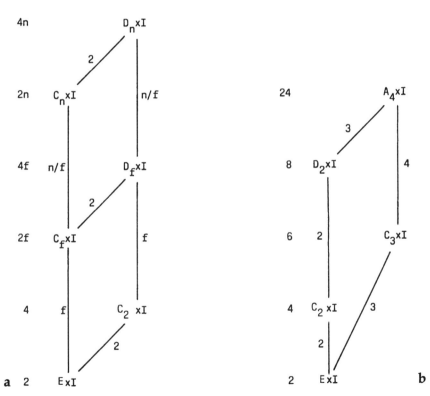

Diagram 1. (a) Dihedral I-subgroups. When n is even and $f = 2$, $C_f \times I = D_1 \times I$. (b) Tetrahedral I-subgroups.

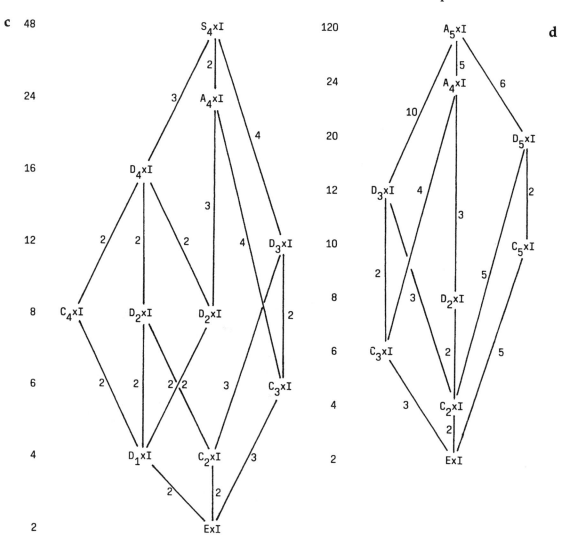

Diagram 1. (c) Octahedral *I*-subgroups. Note distinction of two different subgroups $D_2 \times I$. (d) Icosahedral *I*-subgroups.

Chapter 4   Classification of the Finite Compounds of Cubes

2. $A_4 \times I$

  1. $(2 \mid S_4 \times I \mid A_4 \times I)$
     Higher orientative stability: $(S_4 \times I \mid S_4 \times I)$
     Identical to Section 3.1.1
  2. $(5 \mid A_5 \times I \mid A_4 \times I)$
     Fundamental versatility: 1 position, $2 \mapsto 2\ (2\times)$
     $3 \mapsto 3\ (4\times)$
     $4 \mapsto 2$

     Stability: basic

3. $D_4 \times I$

  1. $(n \mid D_{4n} \times I \mid D_4 \times I)$
     ◊ $n = 1$
       Superior orientative stability: $(S_4 \times I \mid S_4 \times I)$
       Identical to Section 3.1.1
     ◊ $n \geq 2$
       Versatility: 2 positions (congruent compound versions)
       Fundamental versatility: 1 position, $2 \mapsto 2\ (n\times)$
       $4 \mapsto 4n$

       Stability: basic
  2. $(3 \mid S_4 \times I \mid D_4 \times I)$
     Fundamental versatility: 2 positions,
     (A) $2 \mapsto 4\ (2\times)$ (B) see Section 1.1.1
     $4 \mapsto 2\ (2\times)$

     (A) Stability: basic
     (B) Higher stability: $S_4 \times I \mid S_4 \times I$

4. $D_3 \times I$

  1. $(n \mid D_{3n} \times I \mid D_3 \times I)$
     ◊ $n = 1$
       Superior orientative stability: $(S_4 \times I \mid S_4 \times I)$
       Identical to Section 3.1.1

◊ $n \geq 2$

Versatility: 2 positions (congruent compound versions)

Fundamental versatility: 1 position, $2 \mapsto 2\,(n\times)$
$3 \mapsto 3n$

Stability: basic

2. $(4 \mid S_4 \times I \,/\, D_3 \times I)$

Fundamental versatility: 2 positions,

(A) $2 \mapsto 2\,(3\times)$ (B) see Section 1.1.1
$3 \mapsto 3$

(A) Stability: basic

(B) Higher stability: $S_4 \times I \,/\, S_4 \times I$

3. $(10 \mid A_5 \times I \,/\, D_3 \times I)$

Fundamental versatility: 2 positions, $2 \mapsto 2\,(3\times)$
$3 \mapsto 3$

Stability: basic

5. $D_2 \times I$

1. $(n \mid D_{2n} \times I \,/\, D_2 \times I)$

◊ $n = 1$

Superior orientative stability: $(S_4 \times I \,/\, S_4 \times I)$
Identical to Section 3.1.1

◊ $n \geq 2$

Fundamental versatility: 2 positions,

(A) $2 \mapsto 2$ (B) $2 \mapsto 2\,(2\times)$
$2 \mapsto 2n$ $4 \mapsto 2\,(2\times)$
$4 \mapsto 2$ $4 \mapsto 2n$

(A) Stability: basic

(B) Superior stability: $D_{4n} \times I \,/\, D_4 \times I$

2. $(3 \mid A_4 \times I \,/\, D_2 \times I)$

Superior orientative stability: $(S_4 \times I \,/\, D_4 \times I)$
Identical to Section 3.1.3

3. $(6 \mid S_4 \times I \,/\, D_2 \times I)$

Fundamental versatility: 2 positions

(A) $2 \mapsto 4$ (2×)   (B) see Section 1.1.1
$4 \mapsto 2$ (2×)
$4 \mapsto 4$
(A) Stability: basic
(B) Higher stability: $S_4 \times I / S_4 \times I$
4. $(15 | A_5 \times I / D_2 \times I)$
Fundamental versatility: 2 positions
(A) $2 \mapsto 4$ (2×)   (B) see Section 3.1.2,
$4 \mapsto 2$ item 2
(A) Stability: basic
(B) Higher stability: $A_5 \times I / A_4 \times I$

3.2. Rotational Freedom

An orientative compound (**G**/**F**), where **F** is a cyclic *I*-subgroup, may be continuously deformed during the full circular rotation of the descriptive about the sole axis of **F**. To register the positions, an arbitrary initial position has to be chosen for the rotational angle $\mu_0 = 0°$. Any position, with regard to the rotational freedom, can now be determined by the addition of the angle μ to the notation:

$$(m | \mathbf{G}/\mathbf{F} | \mu)$$

which refers to a version of $(m | \mathbf{G}/\mathbf{F})$ where the descriptive has rotated through μ about the freedom axis starting from the initial position. With respect to the versatility, μ is restricted to an interval, and fundamental versatility may restrict this interval to a subset. The versatility and a fundamental versatility will be expressed by the values for μ.

To illustrate a fundamental versatility for subgroups in $A_4 \times I$, $S_4 \times I$, and $A_5 \times I$, the axes and mirrors of such a group are intersected with the circumscribed sphere of the descriptive cube, and this intersection is shown from above, along the axis of freedom. The invisible lower half of the sphere is then a central inverse image of the visible upper half. The axes of the group correspond with points, and the mirrors correspond with greater circles on the sphere. The circumscribed circle is the projection of the sphere and is coplanar with a mirror when the freedom axis is two- or fourfold. In that case, the circumscribed circle also corresponds with a mirror. The descriptive is illustrated on this projection by a number of its axes, of which the points, corresponding with the threefold axes, are also identical with the cube's vertices. The illustrated axes of the cube are selected as being distinct with respect to the symmetry under **G**, the distinction of which is then transmitted throughout the compound. The positive sense of rotation about the freedom axis is always counterclockwise.

Any axis of the descriptive which lies on a cone intersects the sphere in a smaller (or greater) circle, corresponding to a full circle of rotation. Each point of this circle corresponds to a value for μ, which indicates the rotated position of the initial cube. A subset of this circle corresponding to a fundamental versatility is illustrated.

1. $C_4 \times I$

1. $(n \mid C_{4n} \times I \mid C_4 \times I)$
   ◊ $n = 1$
     Superior orientative stability: $(S_4 \times I \mid S_4 \times I)$
     Identical to Section 3.1.1
   ◊ $n \geq 2$
     Superior orientative stability: $(D_{4n} \times I \mid D_4 \times I)$
     Identical to Section 3.1.3, item 1
2. $(2n \mid D_{4n} \times I \mid C_4 \times I)$
   Initial position: $2 \mapsto 2\ (2\times)$
   $\phantom{\text{Initial position: }}4 \mapsto 2\ (2\times)$
   $\phantom{\text{Initial position: }}4 \mapsto 4n$
   Versatility: $[0°, 45°/n]$ (congruent compound versions for μ and $45°/n - \mu$)
   Fundamental versatility: $[0°, 22°30'/n]$, positions $4 \mapsto 4n$
   Stability: basic, except for:

Chapter 4 Classification of the Finite Compounds of Cubes

  a. $\mu_0 = 0°$
  - $n = 1$
    Higher superior stability: $S_4 \times I \ / \ S_4 \times I$
  - $n \geq 2$
    Higher stability: $D_{4n} \times I \ / \ D_4 \times I$
  b. $\mu_1 = 22°30'/n$
    Superior stability: $D_{8n} \times I \ / \ D_4 \times I$
3. $(6| \ S_4 \times I \ / \ C_4 \times I)$
   Initial position: see Section 1.1.1
   Fundamental versatility: $[0°, 45°]$, positions $4 \mapsto 4$
   (Fig. 84).
   Stability: basic, except for
   a. $\mu_0 = 0°$
      Higher stability: $S_4 \times I \ / \ S_4 \times I$
   b. $\mu_1 = 45°$
      Higher stability: $S_4 \times I \ / \ D_4 \times I$

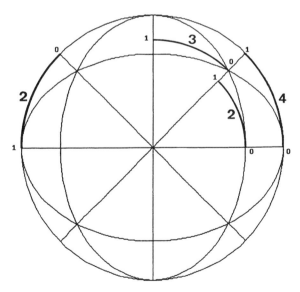

Figure 84. Fundamental versatility of $(6| \ S_4 \times I \ / \ C_4 \times I)$

2. $C_3 \times I$

1. $(n \mid C_{3n} \times I / C_3 \times I)$
   - ◇ $n = 1$
     Superior orientative stability: $(S_4 \times I / S_4 \times I)$
     Identical to Section 3.1.1
   - ◇ $n \geq 2$
     Superior orientative stability: $(D_{3n} \times I / D_3 \times I)$
     Identical to Section 3.1.4, item 1

2. $(2n \mid D_{3n} \times I / C_3 \times I)$
   Initial position: $2 \mapsto 2$ $(3\times)$
   $\qquad\qquad\qquad 3 \mapsto 3n$
   Versatility: $[0°, 60°/n]$ (congruent compound versions for μ and $60°/n - \mu$)
   Fundamental versatility: $[0°, 30°/n]$, positions $3 \mapsto 3$
   Stability: basic, except for
   a. $\mu_0 = 0°$
      Higher stability: $D_{3n} \times I / D_3 \times I$
   b. $\mu_1 = 30°/n$
      Superior stability: $D_{6n} \times I / D_3 \times I$

3. $(4 \mid A_4 \times I / C_3 \times I)$
   Initial position: see Section 1.1.2
   Versatility: $[-60°, 60°]$ (congruent compound versions for $\pm\mu$)
   Fundamental versatility: $[0°, 60°]$, positions $3 \mapsto 3$ (Fig. 85).
   Stability: basic, except for
   a. $\mu_0 = 0°$
      Higher superior stability: $S_4 \times I / S_4 \times I$
   b. $\mu_1 = 60°$
      Superior stability: $S_4 \times I / D_3 \times I$

4. $(8 \mid S_4 \times I / C_3 \times I)$
   Initial position: see Section 1.1.1
   Fundamental versatility: $[0°, 60°]$, positions $3 \mapsto 3$ (Fig. 86)
   Stability: basic, except for
   a. $\mu_0 = 0°$
      Higher stability: $S_4 \times I / S_4 \times I$

Chapter 4   Classification of the Finite Compounds of Cubes   111

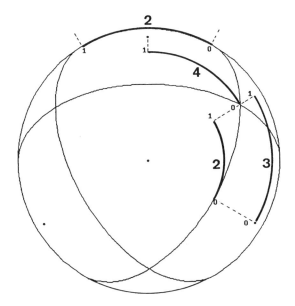

Figure 85. Fundamental versatility of $(4|\ A_4 \times I\ /\ C_3 \times I)$

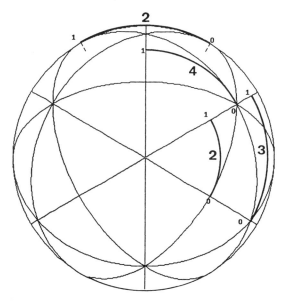

Figure 86. Fundamental versatility of $(8|\ S_4 \times I\ /\ C_3 \times I)$

b. $\mu_1 = 60°$
   Higher stability: $S_4 \times I \mathbin{/} D_3 \times I$
5. $(20 \mathbin{|} A_5 \times I \mathbin{/} C_3 \times I)$
   Initial position: $2 \mapsto 2\ (2\times)$
   $\phantom{\text{Initial position: }}3 \mapsto 3\ (4\times)$
   $\phantom{\text{Initial position: }}4 \mapsto 2$
   Fundamental versatility: $[-22°14'19''.52,\ 37°45'40''.48]$,
   positions $3 \mapsto 3$ (Fig. 87)
   Stability: basic, except for
   a. $\mu_0 = 0°$
      Higher stability: $A_5 \times I \mathbin{/} A_4 \times I$
   b. $\mu_1 = 37°45'40''.48$
      Higher stability: $A_5 \times I \mathbin{/} D_3 \times I$
   c. $\mu_2 = -22°14'19''.52$
      Higher stability: $A_5 \times I \mathbin{/} D_3 \times I$

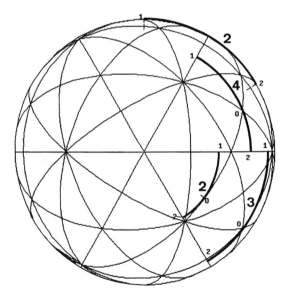

Figure 87. Fundamental versatility of $(20 \mathbin{|} A_5 \times I \mathbin{/} C_3 \times I)$

Chapter 4 Classification of the Finite Compounds of Cubes

3. $C_2 \times I$

1. $(n \mid C_{2n} \times I \, / \, D_1 \times I)$
   - $n = 1$
     Superior orientative stability: $(S_4 \times I \, / \, S_4 \times I)$
     Identical to Section 3.1.1
   - $n \geq 2$
     Superior orientative stability: $(D_{2n} \times I \, / \, D_2 \times I)$
     Identical to Section 3.1.5, item 1
2. $(2n \mid D_{2n} \times I \, / \, D_1 \times I)$
   (A) Positions $2 \mapsto 2n$
     Initial position: $2 \mapsto 2$
     $\phantom{\text{Initial position: }} 4 \mapsto 2$
     Versatility: $[0°, 90°/n]$ (congruent compound versions for μ and $90°/n - \mu$)
     Fundamental versatility: $[0°, 45°/n]$
     Stability: basic, except for
     a. $\mu_0 = 0°$
        - $n = 1$
          Higher superior stability: $S_4 \times I \, / \, S_4 \times I$
        - $n \geq 2$
          Higher stability: $D_{2n} \times I \, / \, D_2 \times I$
     b. $\mu_1 = 35°15'51''.80$
        - $n = 1$
          Superior stability: $D_6 \times I \, / \, D_3 \times I$
     c. $\mu_2 = 45°/n$
          Superior stability: $D_{4n} \times I \, / \, D_2 \times I$
   (B) Positions $4 \mapsto 2n$
     - n is odd
       Superior orientative stability: $(D_{4n} \times I \, / \, C_4 \times I)$
       Identical to Section 3.2.1, item 2
     - $n = 2m$
       Higher orientative stability: $(D_{2n} \times I \, / \, C_4 \times I)$
       $[= (D_{4m} \times I \, / \, C_4 \times I)]$
       Identical to Section 3.2.1, item 2; $n = m$
3. $(n \mid D_n \times I \, / \, C_2 \times I)$
   - $n = 1$. See item 1; $n = 1$

◊ $n = 2$. See item 2; $n = 1$
◊ $n \geq 3$
  (A) positions $2 \mapsto 2$
    ◊ n is odd
      Initial position: $4 \mapsto n$
      Versatility: [0°, 180°] (congruent compound versions for μ and $180° - \mu$)
    ◊ n is even
      Initial position: $2 \mapsto 2$
                    $4 \mapsto 2$ ($2\times$, when $n = 4m$)
                    $4 \mapsto n$
    Fundamental versatility: [0°, 90°]
    Stability: basic, except for
    a. $\mu_0 = 0°$
      ◊ n is odd
        Superior stability: $D_{4n} \times I / D_4 \times I$
      ◊ $n = 2m$, m is odd
        Higher superior stability: $D_{2n} \times I / D_4 \times I$
      ◊ $n = 4k$
        Higher stability: $D_n \times I / D_4 \times I$
    b. $\mu_1 = 54°44'08''$
      ◊ $n = 3m$
        Higher stability: $D_n \times I / D_3 \times I$
      ◊ n and 3 are coprime
        Superior stability: $D_{3n} \times I / D_3 \times I$
    c. $\mu_2 = 90°$
      ◊ n is odd
        Superior stability: $D_{2n} \times I / D_2 \times I$
      ◊ n is even
        Higher stability: $D_n \times I / D_2 \times I$
  (B) positions $4 \mapsto 2$
    ◊ n is odd
      Initial position: $4 \mapsto n$
      Versatility: [0°, 90°] (congruent compound versions for μ and $90° - \mu$)
    ◊ n is even

Chapter 4 Classification of the Finite Compounds of Cubes   115

>   Initial position: $2 \mapsto 2$ ($2\times$ when $n = 4m$)
> $4 \mapsto n$
>   Fundamental versatility: $[0°, 45°]$
>   Stability: basic, except for
>   a. $\mu_0 = 0°$
>   ◇ n is odd
>   Superior stability: $D_{4n} \times I \mid D_4 \times I$
>   ◇ $n = 2m$, m is odd
>   Higher superior stability: $D_{2n} \times I \mid D_4 \times I$
>   ◇ $n = 4k$
>   Higher stability: $D_n \times I \mid D_4 \times I$
>   b. $\mu_1 = 45°$
>   ◇ n is odd
>   Superior stability: $D_{2n} \times I \mid D_2 \times I$
>   ◇ n is even
>   Higher stability: $D_n \times I \mid D_2 \times I$
> 4. $(6 \mid A_4 \times I \mid C_2 \times I)$
>   (A) Positions $2 \mapsto 2$
>   Initial position: $2 \mapsto 2$
> $4 \mapsto 2$
>   Versatility: $[0°, 90°]$ (congruent compound versions for μ and $90° - \mu$)
>   Fundamental versatility: $[0°, 45°]$ (Fig. 88)
>   Stability: basic, except for
>   a. $\mu_0 = 0°$
>   Higher superior stability: $S_4 \times I \mid D_4 \times I$
>   b. $\mu_1 = 45°$
>   Superior stability: $S_4 \times I \mid D_2 \times I$
>   (B) Positions $4 \mapsto 2$
>   Superior orientative stability: $S_4 \times I \mid C_4 \times I$
>   Identical to Section 3.2.1, item 3
> 5. $(12 \mid S_4 \times I \mid D_1 \times I)$
>   Initial position: $2 \mapsto 4$ ($2\times$)
> $4 \mapsto 2$ ($2\times$)
> $4 \mapsto 4$

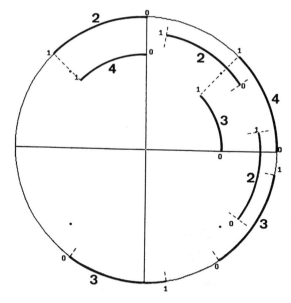

Figure 88. Fundamental versatility of $(6|\ A_4 \times I\ /\ C_2 \times I)$

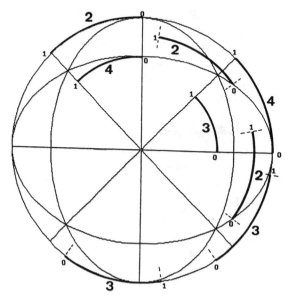

Figure 89. Fundamental versatility of $(12|\ S_4 \times I\ /\ D_1 \times I)$

Fundamental versatility: [0°, 45°], positions (A) 2 ↦ 4
 (B) 4 ↦ 4
(A) Positions 2 ↦ 4 (Fig. 89)
   Stability: basic, except for
   a. $\mu_0 = 0°$
      Higher stability: $S_4 \times I / D_4 \times I$
   b. $\mu_1 = 45°$
      Higher stability: $S_4 \times I / D_2 \times I$
(B) Positions 4 ↦ 4
   Higher orientative stability: $(S_4 \times I / C_4 \times I)$
   Identical to Section 3.2.1, item 3
6. $(12| S_4 \times I / C_2 \times I)$
   (A) Positions 2 ↦ 2
      Initial position: see Section 1.1.1
      Fundamental versatility: [0°, 90°] (Fig. 90)

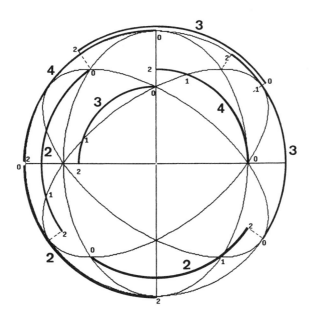

Figure 90. Fundamental versatility of $(12| S_4 \times I / C_2 \times I)$, positions (A)

Stability: basic, except for
a. $\mu_0 = 0°$
Higher stability: $S_4 \times I / S_4 \times I$
b. $\mu_1 = 70°31'43''.60$
Higher stability: $S_4 \times I / D_3 \times I$
c. $\mu_2 = 90°$
Higher stability: $S_4 \times I / D_2 \times I$

(B) Positions $4 \mapsto 2$
Initial position: $2 \mapsto 4$ (2×)
$\phantom{\text{Initial position: }}4 \mapsto 2$
$\phantom{\text{Initial position: }}4 \mapsto 4$
Fundamental versatility: $[0°, 45°]$ (Fig. 91)
Stability: $S_4 \times I / C_2 \times I$, except for
a. $\mu_0 = 0°$
Higher stability: $S_4 \times I / D_4 \times I$
b. $\mu_1 = 45°$
Higher stability: $S_4 \times I / D_2 \times I$

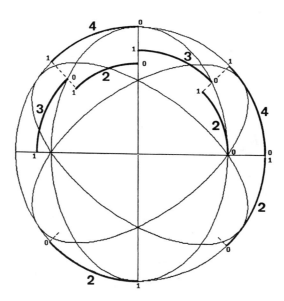

Figure 91. Fundamental versatility of $(12 | S_4 \times I / C_2 \times I)$, positions (B)

Chapter 4 Classification of the Finite Compounds of Cubes

7. $(30 \mid A_5 \times I / C_2 \times I)$
   (A) Positions $2 \mapsto 2$
      Initial position: $2 \mapsto 2$
      $\qquad\qquad\qquad\;\; 4 \mapsto 2$
      Fundamental versatility: $[0°, 90°[$ (Fig. 92)
      Stability: basic, except for
      a. $\mu_0 = 0°$
         Higher stability: $A_5 \times I / D_2 \times I$
      b. $\mu_1 = 14°21'33''.24$
         Higher stability: $A_5 \times I / D_3 \times I$
      c. $\mu_2 = 56°10'10''.36$
         Higher stability: $A_5 \times I / D_3 \times I$
      d. $\mu_3 = 90°$
         Higher stability: $A_5 \times I / D_2 \times I$

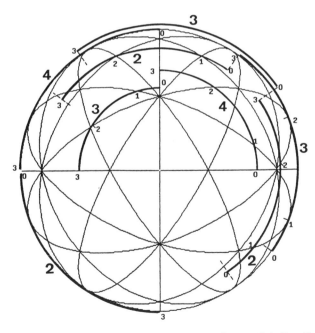

Figure 92. Fundamental versatility of $(30 \mid A_5 \times I / C_2 \times I)$, positions (A)

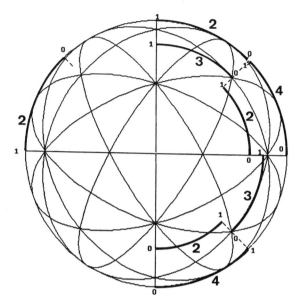

Figure 93. Fundamental versatility of $(30 \mid A_5 \times I / C_2 \times I)$, positions (B)

(B) Positions $4 \mapsto 2$
  Initial position: $3 \mapsto 3$ (4×)
  $4 \mapsto 2$ (2×)
  Fundamental versatility: $[0°, 45°]$ (Fig. 93)
  Stability: basic, except for
  a. $\mu_0 = 0°$
    Higher stability: $A_5 \times I / A_4 \times I$
  b. $\mu_1 = 45°$
    Higher stability: $A_5 \times I / D_2 \times I$

3.3. Central Freedom

The orientative compounds with central freedom are to be considered as the general compounds for the symmetry groups. Besides positions, as mentioned in Sections 3.1 and 3.2, the stability is always $G/E \times I$. Positions with an upgraded stability

Chapter 4 Classification of the Finite Compounds of Cubes

are discussed without going into the details which are found in the previous paragraphs.

1. $(n \mid C_n \times I \mid E \times I)$
   ◊ $n = 1$
     Superior orientative stability: $(S_4 \times I \mid S_4 \times I)$
     Identical to Section 3.1.1
   ◊ $n = 2$
     Stability: basic, except upgraded, when an axis of the cube is coincidental with or perpendicular to the twofold axis of $C_2 \times I$
   ◊ $n \geq 3$
     Stability: basic, except upgraded, when
     ◊ an axis of the cube is coincidental with the n-fold axis of $C_n \times I$ or
     ◊ a two- or fourfold axis of the cube is perpendicular to the n-fold axis of $C_n \times I$
2. $(2n \mid D_n \times I \mid E \times I)$
   ◊ $n = 1$. See Section 3.3.1
   ◊ $n = 2$
     Stability: basic, except upgraded, when
     ◊ an axis of the cube is coincidental with a twofold axis of $D_2 \times I$ or
     ◊ a threefold axis of the cube is coincidental with a threefold axis of $A_4 \times I$, containing $D_2 \times I$ as a subgroup
   ◊ $n \geq 3$
     Stability: basic, except upgraded, when
     ◊ an axis of the cube is coincidental with the n-fold axis of $D_n \times I$ or
     ◊ a two- or fourfold axis of the cube is coincidental with a twofold axis of $D_{2n} \times I$, containing $D_n \times I$ as a subgroup
3. $(12 \mid A_4 \times I \mid E \times I)$
   Stability: basic, except upgraded for positions
   $$2 \mapsto 2$$
   $$4 \mapsto 2$$
   $$\text{or } 3 \mapsto 3$$

4. $(24 \mid S_4 \times I \mid E \times I)$
   Stability: basic, except upgraded for positions

   $$\begin{aligned} 2 &\mapsto 2 \\ 2 &\mapsto 4 \\ 4 &\mapsto 2 \\ 4 &\mapsto 4 \\ \text{or } 3 &\mapsto 3 \end{aligned}$$

5. $(60 \mid A_5 \times I \mid E \times I)$
   Stability: basic, except upgraded for positions

   $$\begin{aligned} 2 &\mapsto 2 \\ 4 &\mapsto 2 \\ \text{or } 3 &\mapsto 3 \end{aligned}$$

3.4. Maximal Orientative Compounds

From the orientative compounds, those with a maximal orientative stability are listed in Table 22.

4. Classification of the Actual Compounds

From the set of maximal orientative compounds, all the distinct versions that are mutually noncongruent are now classified. These are defined as *the versions with a basic stability*.

Let an orientation A contain a cube position p that has a stabilizer $\mathbf{H}_p$, where $\mathbf{F} \subset \mathbf{H}_p$ is proper. Because $\mathbf{H}_p$ is also in one of the nine I-subgroup classes of $S_4 \times I$, $\mathbf{H}_p$ determines a second orientation B such that $B \subset A$ is a proper subset. Hence, such a position is found back with another orientative compound, where it has a basic stability. Alternatively, let p have a stabilizer in a higher I-group. Also then, the position is found back similarly but under a higher I-group.

The notation of a compound is now based on that of an orientative compound, but more specific, and provided with some additional classificatory information:

Chapter 4 Classification of the Finite Compounds of Cubes

Table 22. The orientative compounds with maximal orientative stability.

Group	Rigid	Rotational Freedom	Central Freedom
Cyclic			
$n \geq 2$			$(n\mid C_n \times I\,/\,E \times I)$
Dihedral			
$n \geq 1$		$(2n\mid D_{4n} \times I\,/\,C_{4n} \times I)$	
		$(2n\mid D_{3n} \times I\,/\,C_{3n} \times I)$	
		$(2n\mid D_{2n} \times I\,/\,D_1 \times I)$	
			$(2n\mid D_n \times I\,/\,E \times I)$
$n \geq 2$	$(n\mid D_{4n} \times I\,/\,D_4 \times I)$		
	$(n\mid D_{3n} \times I\,/\,D_3 \times I)$		
	$(n\mid D_{2n} \times I\,/\,D_2 \times I)$		
$n \geq 3$		$(n\mid D_n \times I\,/\,C_2 \times I)$	
Tetrahedral		$(4\mid A_4 \times I\,/\,C_3 \times I)$	
		$(6\mid A_4 \times I\,/\,C_2 \times I)$	
			$(12\mid A_4 \times I\,/\,E \times I)$
Octahedral	$(1\mid S_4 \times I\,/\,S_4 \times I)$		
	$(3\mid S_4 \times I\,/\,D_4 \times I)$		
	$(4\mid S_4 \times I\,/\,D_3 \times I)$		
	$(6\mid S_4 \times I\,/\,D_2 \times I)$		
		$(6\mid S_4 \times I\,/\,C_4 \times I)$	
		$(8\mid S_4 \times I\,/\,C_3 \times I)$	
		$(12\mid S_4 \times I\,/\,D_1 \times I)$	
		$(12\mid S_4 \times I\,/\,C_2 \times I)$	
			$(24\mid S_4 \times I\,/\,E \times I)$
Icosahedral	$(5\mid A_5 \times I\,/\,A_4 \times I)$		
	$(10\mid A_5 \times I\,/\,D_3 \times I)$		
	$(15\mid A_5 \times I\,/\,D_2 \times I)$		
		$(20\mid A_5 \times I\,/\,C_3 \times I)$	
		$(30\mid A_5 \times I\,/\,C_2 \times I)$	
			$(60\mid A_5 \times I\,/\,E \times I)$

$mX|\mathbf{G}/\mathbf{H}|Y$

G: symmetry group
H: stabilizer
G/H: stability
m: number of distinct constituents
X: descriptive position A or B, according to the positionary specification, provided with the orientative compound in Section (with rotational freedom only). X is deleted when no ambiguity occurs.
Y: numbered version A or B (with finite freedom), or angle μ of rotation, restricted to described compounds in the fundamental versatility (with rotational freedom).* In either case, Y may be deleted when such information is irrelevant.

4.1. Cyclic

$n|\ C_n \times I\ /\ E \times I$

$n \geq 2$
Positions: Centered.
Freedom: Central, limited to positions, where
 ◊ $n = 2$:
  an axis of the cube is not coincidental with, nor perpendicular to, the twofold axis of $C_2 \times I$
 ◊ $n \geq 3$:
  ◊ an axis of the cube is not coincidental with the n-fold axis of $C_n \times I$ nor
  ◊ is a two- or fourfold axis of the cube perpendicular to the n-fold axis of $C_n \times I$

When $n = 2$, the compound is composed of the cube and its reflected image in the sole mirror of $C_2 \times I$.

---

* Fundamental versatility is also here defined as a set of distinct compound versions that are mutually noncongruent.

4.2. Dihedral

1. $n \mid D_{4n} \times I \mid D_4 \times I$

$n \geq 2$
Positions: $4 \mapsto 4n$
$4 \mapsto 2\ (2\times)$
$2 \mapsto 2\ (2\times)$
Freedom: Rigid
Fundamental versatility: 1 position
Orientative compound: $(n \mid D_{4n} \times I \mid D_4 \times I)$
A view of this dihedral compound along the $4n$-fold axis reveals n squares, as described in a regular polygon $\{4n\}$.
The compound of two cubes in Fig. 83 illustrates $2 \mid D_8 \times I \mid D_4 \times I$.

2. $n \mid D_{3n} \times I \mid D_3 \times I$

$n \geq 2$
Positions: $3 \mapsto 3n$
$2 \mapsto 2\ (3\times)$
Freedom: Rigid
Fundamental versatility: 1 position
Orientative compound: $(n \mid D_{3n} \times I \mid D_3 \times I)$
The compound in Fig. 81 illustrates $2 \mid D_6 \times I \mid D_3 \times I$.

3. $n \mid D_{2n} \times I \mid D_2 \times I$

$n \geq 2$
Positions: $2 \mapsto 2n$
$4 \mapsto 2$
$2 \mapsto 2$
Freedom: Rigid
Fundamental versatility: 1 position
Orientative compound: $(n \mid D_{2n} \times I \mid D_2 \times I)$, being in version
(B): $n \mid D_{4n} \times I \mid D_4 \times I$

A view of this compound along the $2n$-fold axis reveals n edges as the central diagonals in a regular polygon $\{2n\}$. The compound sum in Fig. 81 is the sum $2|\ D_4 \times I\ /\ D_2 \times I\ +\ 2|\ D_8 \times I\ /\ D_4 \times I$, whereas Fig. 94 shows the compound when $n = 3$.

4. $2n|\ D_{4n} \times I\ /\ C_4 \times I$

Positions: $4 \mapsto 4n$
Freedom: Rotational
Fundamental versatility: $]\mu_0, \mu_1[$ (see Section 3.2.1, item 2)
Orientative compound: $(2n|\ D_{4n} \times I\ /\ C_4 \times I\ |\ \mu)$, being in the end versions (corresponding with μ_i):

$\mu_0 = 0°$: $1|\ S_4 \times I\ /\ S_4 \times I$ (when $n = 1$)
               $n|\ D_{4n} \times I\ /\ D_4 \times I$ (when $n \geq 2$)

$\mu_1 = 22°30'/n$: $2n|\ D_{8n} \times I\ /\ D_4 \times I$

A version of this is seen in Fig. 95, when $n = 1$.
When $n = 1$, a pair of cubes shares a fourfold axis with $D_4 \times I$, and when $n \geq 2$, a pair of $n|\ D_{4n} \times I\ /\ D_4 \times I$ shares a $4n$-fold axis with $D_{4n} \times I$. Each of these rotates in a different sense with respect to the rotational freedom. The compound can also be considered as the product of an n-fold subcompound $2|\ D_4 \times I\ /\ C_4 \times I$, the cubes of which rotate in opposite senses.

5. $2n|\ D_{3n} \times I\ /\ C_3 \times I$

Positions: $3 \mapsto 3n$
Freedom: Rotational
Fundamental versatility: $]\mu_0, \mu_1[$ (see Section 3.2.2, item 2)
Orientative compound: $(2n|\ D_{3n} \times I\ /\ C_3 \times I\ |\ \mu)$, being in the end versions (corresponding with μ_i):

$\mu_0 = 0°$: $1|\ S_4 \times I\ /\ S_4 \times I$ (when $n = 1$)
               $n|\ D_{3n} \times I\ /\ D_3 \times I$ (when $n \geq 2$)

$\mu_1 = 30°/n$: $2n|\ D_{6n} \times I\ /\ D_3 \times I$

Chapter 4 Classification of the Finite Compounds of Cubes

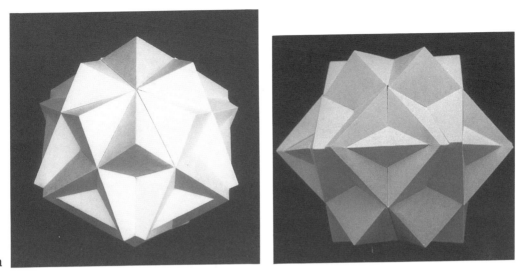

Figure 94. $3|\ D_6 \times I\ /\ D_2 \times I$: (a) general view from a point in a mirror; (b) along a twofold axis

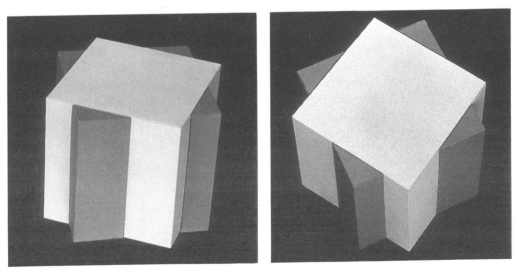

Figure 95. $2|\ D_4 \times I\ /\ C_4 \times I\ |15°$: (a) general view from a point in a mirror; (b) the same as (a) but from a different point of view

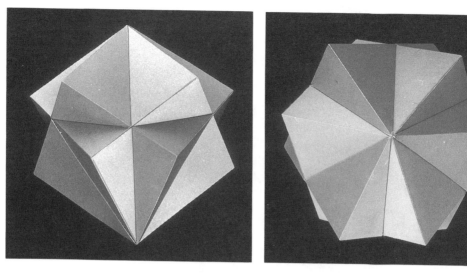

Figure 96. $2| D_3 \times I / C_3 \times I | 22°14'19''.52$: (a) general view; (b) along a threefold axis

A version of this is seen in Fig. 96, when $n = 1$.

When $n = 1$, a pair of cubes shares a threefold axis with $D_3 \times I$, and when $n \geq 2$, a pair of $n| D_{3n} \times I / D_3 \times I$ shares the $2n$-fold axis with $D_{3n} \times I$. Each of these rotate in a different sense with respect to the rotational freedom. The compound can also be considered as the product of an n-fold subcompound $2| D_3 \times I / C_3 \times I$, the cubes of which rotate in opposite senses.

6. $2n| D_{2n} \times I / D_1 \times I$

   Positions: $2 \mapsto 2n$
   Freedom: Rotational
   ◊ $n = 1$
     Fundamental versatility: $]\mu_0, \mu_1[\cup]\mu_1, \mu_2[$ [see Section 3.2.3, item 2(A)]
     Orientative compound: $(2| D_2 \times I / D_1 \times I |\mu)$, being in the end versions (corresponding with μ_i):

$\mu_0 = 0°$: $1|\ S_4 \times I\ /\ S_4 \times I$
$\mu_1 = 35°15'51''.80$: $2|\ D_6 \times I\ /\ D_3 \times I$
$\mu_2 = 45°$: $2|\ D_4 \times I\ /\ D_2 \times I$

Such a version is seen in Fig. 97.

◊ $n \geq 2$

Fundamental versatility: $]\mu_0, \mu_2[$ [see Section 3.2.3, item 2(A)]

Orientative compound: $(2n|\ D_{2n} \times I\ /\ D_1 \times I\ |\mu)$, being in the end versions (corresponding with μ_i):

$\mu_0 = 0°$: $n|\ D_{2n} \times I\ /\ D_2 \times I$
$\mu_2 = 45°/n$: $2n|\ D_{4n} \times I\ /\ D_2 \times I$

When $n = 1$, a pair of cubes shares a twofold axis with $D_2 \times I$, and when $n \geq 2$, a pair of $n|\ D_{2n} \times I\ /\ D_2 \times I$ shares the $2n$-fold axis with $D_{2n} \times I$. Each of these rotates in a different sense with respect to the rotational freedom. The compound can also be considered as the product of an n-fold subcompound $2|\ D_2 \times I\ /\ D_1 \times I$, the cubes of which rotate in opposite senses.

7. $nA|\ D_n \times I\ /\ C_2 \times I$

$n \geq 3$
Positions: $2 \mapsto 2$
Freedom: Rotational
Fundamental versatility: $]\mu_0, \mu_1[\ \cup\ ]\mu_1, \mu_2[$ [see Section 3.2.3, item 3(A)]
Orientative compound: $(n|\ D_n \times I\ /\ C_2 \times I\ |\mu)$, being in the end versions (corresponding with μ_i):

$\mu_0 = 0°$
  ◊ n is odd: $n|\ D_{4n} \times I\ /\ D_4 \times I$
  ◊ $n = 2m$, m is odd: $n/2|\ D_{2n} \times I\ /\ D_4 \times I$
  ◊ $n = 4m$: $n/4|\ D_n \times I\ /\ D_4 \times I$

$\mu_1 = 54°44'08''.20$
  ◊ n is coprime with 3: $n|\ D_{3n} \times I\ /\ D_3 \times I$
  ◊ $n = 3m$: $n/3|\ D_n \times I\ /\ D_3 \times I$

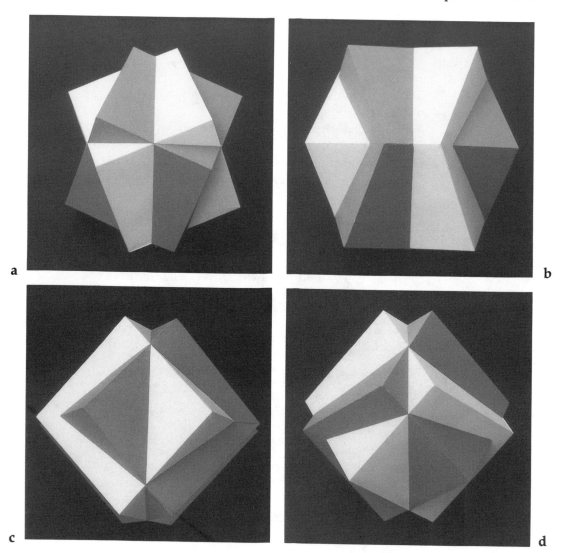

Figure 97. $2 \mid D_2 \times I \mid D_1 \times I \mid 22°30'$: (a) along the shared twofold axis; (b) along a second twofold axis; (c) along the third twofold axis; (d) general view from a point in a mirror

$\mu_2 = 90°$
◊ n is odd: $n|\, D_{2n} \times I\, /\, D_2 \times I$
◊ n is even: $n/2|\, D_n \times I\, /\, D_2 \times I$

When n is odd, $D_n \times I$ shares each twofold axis with a single cube. Cubes about two successive axes rotate in opposite senses with respect to the rotational freedom. This is easily understood, as there is a mirror of $D_n \times I$ acting like a bisector plane of two successive twofold axes. When $n = 2m$, the compound is the product $2 \times mA|\, D_m \times I\, /\, C_2 \times I$, i.e., a pair of odd compounds of this type, rotated images of one another through $180°/m$ about the n-fold axis. Clearly, the pairs are not coincidental as successive cubes have rotated in opposite senses with respect to the rotational freedom. The composition of m pairs about half the set of twofold axes is also obvious by observing the compound as being the product $m \times 2|\, D_2 \times I\, /\, C_2 \times I$. The two cubes in a pair rotate in opposite senses with respect to the rotational freedom, due to a mirror of $D_n \times I$ containing the axis. The only exception for the latter product is for $\mu = 45°$, where the subcompound has a dihedral symmetry (see also Chapter 5):

$$nA|\, D_n \times I\, /\, C_2 \times I\, |45° = m \times 2|\, D_4 \times I\, /\, D_2 \times I$$

8. $nB|\, D_n \times I\, /\, C_2 \times I$

$n \geq 3$
Positions: $4 \mapsto 2$
Freedom: Rotational
Fundamental versatility: $]\mu_0, \mu_1[$ [see Section 3.2.3, item 3(B)]
Orientative compound: $(n|\, D_n \times I\, /\, C_2 \times I\, |\mu)$, being in the end versions (corresponding with μ_i):

$\mu_0 = 0°$
◊ n is odd: $n|\, D_{4n} \times I\, /\, D_4 \times I$
◊ $n = 2m$, m is odd: $n/2|\, D_{2n} \times I\, /\, D_4 \times I$
◊ $n = 4m$: $n/4|\, D_n \times I\, /\, D_4 \times I$

$\mu_1 = 45°$
- ◇ n is odd: $n \mid D_{2n} \times I \mathbin{/} D_2 \times I$
- ◇ n is even: $n/2 \mid D_n \times I \mathbin{/} D_2 \times I$

This compound is the twin of compound 7, based on the same orientative compound. Hence, identical conclusions are valid here, with the difference that a fourfold axis of the descriptive is now coincidental with a twofold axis of $D_n \times I$. When $n = 2m$, the exception here is for 22°30′:

$$nB \mid D_n \times I \mathbin{/} C_2 \times I \mid 22°30' = m \times 2 \mid D_8 \times I \mathbin{/} D_4 \times I$$

9. $2n \mid D_n \times I \mathbin{/} E \times I$

$n \geq 2$
Positions: Centered
Freedom: Central, limited to positions, where
- ◇ $n = 2$:
  - ◇ an axis of the cube is not coincidental with a twofold axis of $D_2 \times I$ nor
  - ◇ is a threefold axis of the cube coincidental with a threefold axis of $A_4 \times I$, containing $D_2 \times I$ as a subgroup
- ◇ $n \geq 3$
  - ◇ an axis of the cube is not coincidental with the n-fold axis of $D_n \times I$ nor
  - ◇ is a two- or fourfold axis of the cube coincidental with a twofold axis of $D_{2n} \times I$, containing $D_n \times I$ as a subgroup

4.3. Tetrahedral

1. $4 \mid A_4 \times I \mathbin{/} C_3 \times I$

    Positions: $3 \mapsto 3$
    Freedom: Rotational
    Fundamental versatility: $]\mu_0, \mu_1[$ (see Section 3.2.2, item 3)

Chapter 4   Classification of the Finite Compounds of Cubes   133

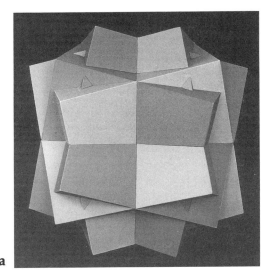
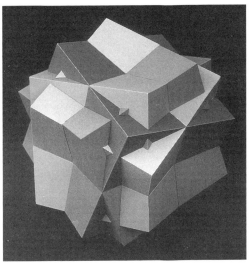

Figure 98. 4| $A_4 \times I$ / $C_3 \times I$ | 22°14′19″.52: (a) along a twofold axis; (b) along a threefold axis

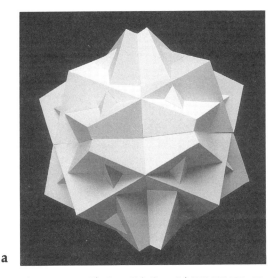
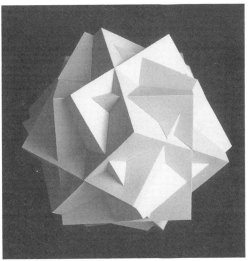

Figure 99. 4| $A_4 \times I$ / $C_3 \times I$ | 37°45′40″.48: (a) along a twofold axis; (b) along a threefold axis

Orientative compound: $(4| A_4 \times I / C_3 \times I |\mu)$, being in the end versions (corresponding with μ_i):

$$\mu_0 = 0°: \quad 1| S_4 \times I / S_4 \times I$$
$$\mu_1 = 60°: \quad 4| S_4 \times I / D_3 \times I$$

Each of the four threefold axes of $A_4 \times I$ is shared with a single cube. With regard to the rotational freedom, cubes about threefold axes in adjacent quadrants rotate in opposite senses because the bisector plane of these axes is a mirror of $A_4 \times I$. Two versions are seen in Figs. 98 and 99.

2. $6| A_4 \times I / C_2 \times I$

Positions: $2 \mapsto 2$
Freedom: Rotational
Fundamental versatility: $]\mu_0, \mu_1[$ (see Section 3.2.3, item 4)
Orientative compound: $(6| A_4 \times I / C_2 \times I |\mu)$, being in the end versions (corresponding with μ_i):

$$\mu_0 = 0°: \quad 3| S_4 \times I / D_4 \times I \text{ (see Section 4.4.2)}$$
$$\mu_1 = 45°: \quad 6| S_4 \times I / D_2 \times I \text{ (see Section 4.4.4)}$$

A pair of cubes is sharing a twofold axis with $A_4 \times I$. The cubes of a pair rotate in opposite senses with respect to the rotational freedom. Two versions are seen in Figs. 100 and 101.

3. $12| A_4 \times I / E \times I$

Positions: Centered
Freedom: Central. Positions, distinct from either

$$2 \mapsto 2$$
$$4 \mapsto 2$$
$$\text{or } 3 \mapsto 3$$

Chapter 4 Classification of the Finite Compounds of Cubes

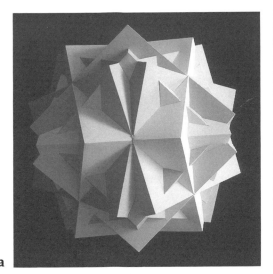
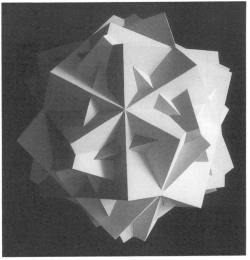

Figure 100. $2 \mid A_4 \times I \mathbin{/} C_2 \times I \mid 14°21'33''.24$: (a) along a twofold axis; (b) along a threefold axis

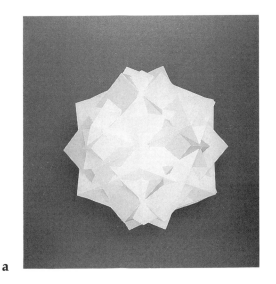

Figure 101. $2 \mid A_4 \times I \mathbin{/} C_2 \times I \mid 22°30'$: (a) along a twofold axis; (b) along a threefold axis

4.4. Octahedral

1. $1 \mid S_4 \times I \mid S_4 \times I$

Positions: $2 \mapsto 2$ (6×)
$3 \mapsto 3$ (4×)
$4 \mapsto 4$ (3×)
Freedom: Rigid
Fundamental versatility: 1 position
Orientative compound: $(1 \mid S_4 \times I \mid S_4 \times I)$
This is the *trivial* compound with one constituent only.

2. $3 \mid S_4 \times I \mid D_4 \times I$

Positions: $4 \mapsto 4$
$4 \mapsto 2$ (2×)
$2 \mapsto 4$ (2×)
Freedom: Rigid
Fundamental versatility: 1 position
Orientative compound: $(3 \mid S_4 \times I \mid D_4 \times I)$, being in version (B): $1 \mid S_4 \times I \mid S_4 \times I$
See Fig. 102.

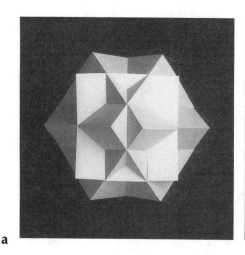
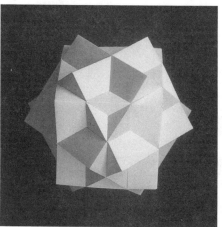

Figure 102. $3 \mid S_4 \times I \mid D_4 \times I$, the "classic" uniform compound of three cubes: (a) along a twofold axis; (b) along a threefold axis

3. $4 | S_4 \times I / D_3 \times I$

>Positions: $3 \mapsto 3$
>$\qquad\quad\;\; 2 \mapsto 2\ (3\times)$
>Freedom: Rigid
>Fundamental versatility: 1 position
>Orientative compound: $(4| S_4 \times I / D_3 \times I)$, being in version
>   (B): $1 | S_4 \times I / S_4 \times I$
>See Fig. 103.

4. $6 | S_4 \times I / D_2 \times I$

>Positions: $4 \mapsto 2$
>$\qquad\quad\;\; 2 \mapsto 4$
>$\qquad\quad\;\; 2 \mapsto 2$
>Freedom: Rigid
>Fundamental versatility: 1 position
>Orientative compound: $(6| S_4 \times I / D_2 \times I)$, being in version
>   (B): $1 | S_4 \times I / S_4 \times I$
>See Fig. 104.

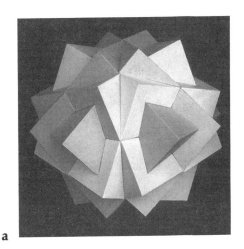
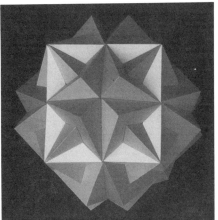

a b

Figure 103. $4 | S_4 \times I / D_3 \times I$, Bakos's compound: (a) along a two-fold axis; (b) along a fourfold axis

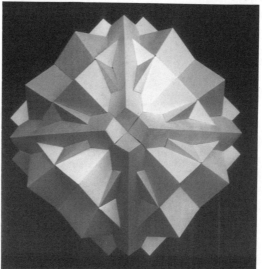

Figure 104. $6| S_4 \times I / D_2 \times I$: (a) along a twofold axis; (b) along a threefold axis; (c) along a fourfold axis

5. $6| S_4 \times I / C_4 \times I$

Positions: $4 \mapsto 4$
Freedom: Rotational
Fundamental versatility: $]\mu_0, \mu_1[$ (see Section 3.2.1, item 3)
Orientative compound: $(6| S_4 \times I / C_4 \times I |\mu)$, being in the end versions (corresponding with μ_i):

$$\mu_0 = 0°: \quad 1| S_4 \times I / S_4 \times I$$
$$\mu_1 = 45°: \quad 3| S_4 \times I / D_4 \times I$$

A pair of cubes shares a fourfold axis with $S_4 \times I$ (Fig. 105). The cubes rotate in opposite senses with respect to the rotational freedom.

Figure 105. $6| S_4 \times I / C_4 \times I$, Skilling's uniform compound with rotational freedom, seen from a point in a mirror

6. $8 | S_4 \times I / C_3 \times I$

Positions: $3 \mapsto 3$
Freedom: Rotational
Fundamental versatility: $]\mu_0, \mu_1[$ (see Section 3.2.2, item 4)
Orientative compound: $(8 | S_4 \times I / C_3 \times I | \mu)$, being in the end versions (corresponding with μ_i):

$\mu_0 = 0°:$ $1 | S_4 \times I / S_4 \times I$
$\mu_1 = 60°:$ $4 | S_4 \times I / D_3 \times I$

A pair of cubes shares a threefold axis with $S_4 \times I$ (Fig. 106). The cubes rotate in opposite senses with respect to the rotational freedom.

7. $12 | S_4 \times I / D_1 \times I$

Positions: $2 \mapsto 4$
Freedom: Rotational
Fundamental versatility: $]\mu_0, \mu_1[$ [see Section 3.2.3, item 5(A)]
Orientative compound: $(12 | S_4 \times I / D_1 \times I | \mu)$, being in the end versions (corresponding with μ_i):

$\mu_0 = 0°:$ $3 | S_4 \times I / D_4 \times I$
$\mu_1 = 45°:$ $6 | S_4 \times I / D_2 \times I$

Four cubes share a twofold axis, which is coincidental with a fourfold axis of $S_4 \times I$. The four cubes consist of two pairs, $2 | D_2 \times I / C_2 \times I$, being images of one another through a rotation of 90° about the common axis. The cubes of each such pair rotate in opposite senses, with respect to the rotational freedom. The four cubes consists also of a pair of compounds, $2 | D_4 \times I / D_2 \times I$. These pairs rotate in opposite senses with respect to the rotational freedom.

Chapter 4 Classification of the Finite Compounds of Cubes

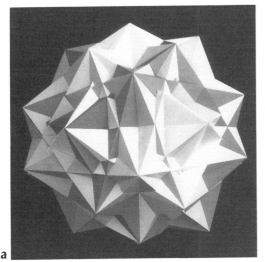

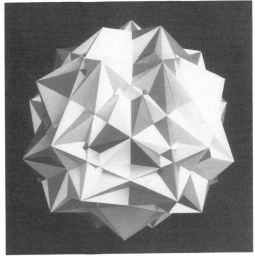

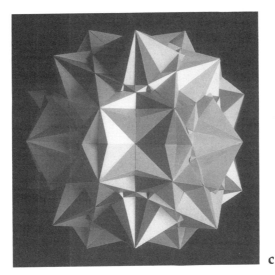

Figure 106. $8 \mid S_4 \times I / C_3 \times I$: (a) along a twofold axis; (b) along a threefold axis; (c) along a fourfold axis

8. $12A \mid S_4 \times I / C_2 \times I$

Positions: $2 \mapsto 2$
Freedom: Rotational
Fundamental versatility: $]\mu_0, \mu_1[\; \cup \;]\mu_1, \mu_2[$ [see Section 3.2.3, item 6(A)]
Orientative compound: $(12 \mid S_4 \times I / C_2 \times I \mid \mu)$, being in the end versions (corresponding with μ_i):

$\mu_0 = 0°$: $1 \mid S_4 \times I / S_4 \times I$
$\mu_1 = 70°31'43''.60$: $4 \mid S_4 \times I / D_3 \times I$
$\mu_2 = 90°$: $6 \mid S_4 \times I / D_2 \times I$

A pair of cubes shares a twofold axis with $S_4 \times I$. The cubes rotate in opposite senses with respect to the rotational freedom.

9. $12B \mid S_4 \times I / C_2 \times I$

Positions: $4 \mapsto 2$
Freedom: Rotational
Fundamental versatility: $]\mu_0, \mu_1[$ [see Section 3.2.3, item 6(B)]
Orientative compound: $(12 \mid S_4 \times I / C_2 \times I \mid \mu)$, being in the end versions (corresponding with μ_i):

$\mu_0 = 0°$: $3 \mid S_4 \times I / D_4 \times I$
$\mu_1 = 45°$: $6 \mid S_4 \times I / D_2 \times I$

This is a twin of compound 8, based on the same orientative compound. A pair of cubes share a fourfold axis, which is coincidental with a twofold axis of $S_4 \times I$ (Fig. 107). The cubes rotate in opposite senses with respect to the rotational freedom.

Chapter 4 Classification of the Finite Compounds of Cubes

Figure 107. $12B \mid S_4 \times I / C_2 \times I$: (a) along a twofold axis; (b) along a threefold axis; (c) along a fourfold axis

10. $24| S_4 \times I / E \times I$

Positions: Centered
Freedom: Central. Positions, distinct from either

$$\begin{aligned} 2 &\mapsto 2 \\ 2 &\mapsto 4 \\ 4 &\mapsto 2 \\ 4 &\mapsto 4 \\ \text{or } 3 &\mapsto 3 \end{aligned}$$

4.5. Icosahedral

1. $5| A_5 \times I / A_4 \times I$

    Positions: $3 \mapsto 3$ (4×)
    $\phantom{\text{Positions: }}4 \mapsto 2$ (3×)
    Freedom: Rigid
    Fundamental versatility: 1 position
    Orientative compound: $(5| A_5 \times I / A_4 \times I)$
    This compound is the classic compound of five cubes (Fig. 108) (see the Historical Appendix). $5| A_5 \times I / A_4 \times I$ can be described in a regular dodecahedron, i.e., its *casing* is a dodecahedron.

2. $10| A_5 \times I / D_3 \times I$

    Positions: $3 \mapsto 3$
    $\phantom{\text{Positions: }}2 \mapsto 2$ (3×)
    Freedom: Rigid
    Fundamental versatility: 2 positions (A and B, see further specifications for proper distinction)
    Orientative compound: $(10| A_5 \times I / D_3 \times I)$
    See Figs. 109–110.

Chapter 4 Classification of the Finite Compounds of Cubes

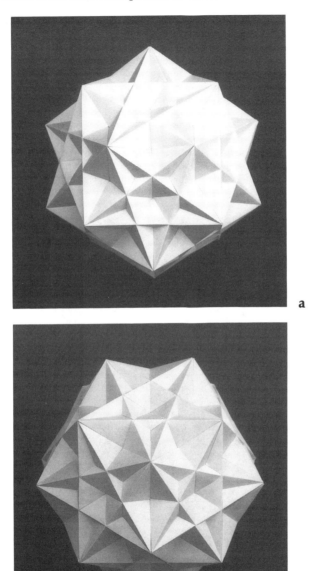

Figure 108. $5 \mid A_5 \times I / A_4 \times I$, the "classic," uniform compound of five cubes: (a) along a twofold axis; (b) along a threefold axis

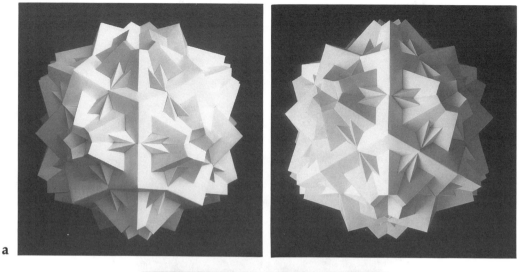
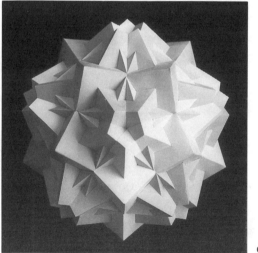

Figure 109. $10 \mid A_5 \times I \mid D_3 \times I \mid A$: (a) along a twofold axis; (b) along a threefold axis; (c) along a fivefold axis

Chapter 4 Classification of the Finite Compounds of Cubes

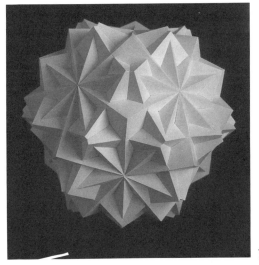

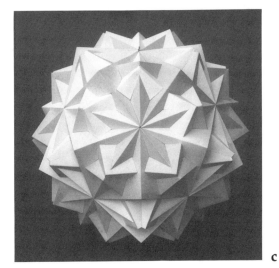

Figure 110. $10 | A_5 \times I \ / \ D_3 \times I \ | B$: (a) along a twofold axis; (b) along a threefold axis; (c) along a fivefold axis

3. $15 | A_5 \times I / D_2 \times I$

Positions: $4 \mapsto 2$
$2 \mapsto 2 \ (2\times)$
Freedom: Rigid
Fundamental versatility: 1 position
Orientative compound: $(15 | A_5 \times I / D_2 \times I)$, being in version (B): $5 | A_5 \times I / A_4 \times I$
See Fig. 111.

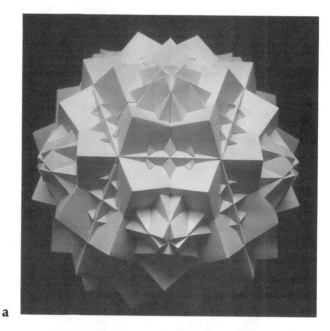

a

Figure 111. $15 | A_5 \times I / D_2 \times I$: (a) along a twofold axis; (b) along a threefold axis; (c) along a fivefold axis

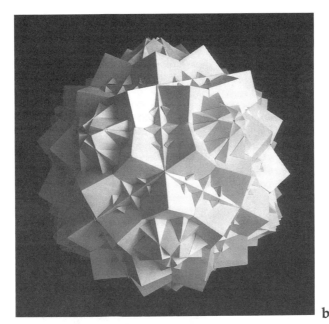

b

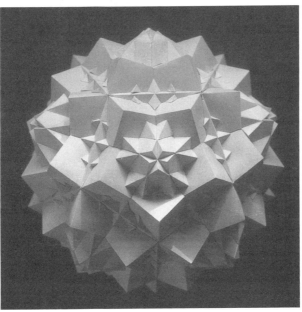

c

4. $20 \mid A_5 \times I \mid C_3 \times I$

Positions: $3 \mapsto 3$
Freedom: Rotational
Fundamental versatility: $]\mu_2, \mu_0[\cup]\mu_0, \mu_1[$ (see Section 3.2.2, item 5)
Orientative compound: $(20 \mid A_5 \times I \mid C_3 \times I \mid \mu)$, being in the end versions (corresponding with μ_i):

$\mu_0 = 0°$: $\quad\quad\quad 5 \mid A_5 \times I \mid A_4 \times I$
$\mu_1 = 37°45'40''.48$: $\quad 10 \mid A_5 \times I \mid D_3 \times I \mid A$
$\quad\quad\quad\quad\quad\quad\quad\quad\quad$ (see also compound 5)
$\mu_2 = -22°14'19''.52$: $\quad 10 \mid A_5 \times I \mid D_3 \times I \mid B$

Note in Fig. 110 how in version B a threefold axis of the descriptive is extremely close to a twofold axis of $A_5 \times I$. The arc is only $1°26'02''.16$ (see also the Historical Appendix).

A pair of cubes shares a threefold axis with $A_5 \times I$ (Figs. 112–113). The cubes rotate in opposite senses with respect to the rotational freedom.

5. $30A \mid A_5 \times I \mid C_2 \times I$

Positions: $2 \mapsto 2$
Freedom: Rotational
Fundamental versatility: $]\mu_0, \mu_1[\cup]\mu_1, \mu_2[\cup]\mu_2, \mu_3[$ [see Section 3.2.3, item 7(A)]
Orientative compound: $(30 \mid A_5 \times I \mid C_3 \times I \mid \mu)$, being in the end versions (corresponding with μ_i):

$\mu_0 = 0°$: $\quad\quad\quad 15 \mid A_5 \times I \mid D_2 \times I$
$\mu_1 = 14°21'33''.24$: $\quad 10 \mid A_5 \times I \mid D_3 \times I \mid A$
$\mu_2 = 56°10'10''.36$: $\quad 10 \mid A_5 \times I \mid D_3 \times I \mid B$
$\mu_3 = 90°$: $\quad\quad\quad 15 \mid A_5 \times I \mid D_2 \times I$

A pair of cubes shares a twofold axis with $A_5 \times I$. The cubes rotate in opposite senses with respect to the rotational freedom.

Chapter 4 Classification of the Finite Compounds of Cubes   151

6. $30B \mid A_5 \times I \mid C_2 \times I$

   Positions: $4 \mapsto 2$
   Freedom: Rotational
   Fundamental versatility: $]\mu_0, \mu_1[$ [see Section 3.2.3, item 7(B)]
   Orientative compound: $(30 \mid A_5 \times I \mid C_2 \times I \mid \mu)$, being in the end versions (corresponding with μ_i):

   $$\mu_0 = 0°: \quad 5 \mid A_5 \times I \mid A_4 \times I$$
   $$\mu_1 = 45°: \quad 15 \mid A_5 \times I \mid D_2 \times I$$

   This is a twin of compound 5, based on the same orientative compound. A pair of cubes shares a fourfold axis, which is coincidental with a twofold axis of $A_5 \times I$. The cubes rotate in opposite senses with respect to the rotational freedom.

7. $60 \mid A_5 \times I \mid E \times I$

   Positions: Centered
   Freedom: Central. Positions, distinct from either

   $$2 \mapsto 2$$
   $$4 \mapsto 2$$
   $$\text{or } 3 \mapsto 3$$

Figure 112. 20| $A_5 \times I$ / $C_3 \times I$ | 14°05′32″.65: (a) along a twofold axis; (b) along a threefold axis; (c) along a fivefold axis

b

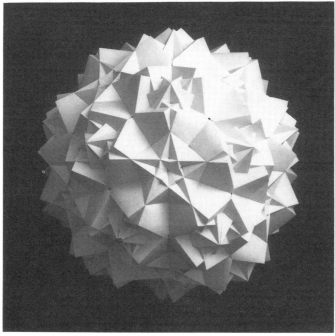

c

Part II Compounds of Cubes

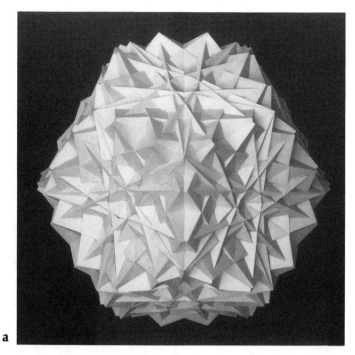

a

Figure 113. 20| $A_5 \times I$ / $C_3 \times I$ | $-12°22'36''$.50: (a) general view from a point in a mirror; (b) along a twofold axis; (c) along a fivefold axis

Chapter 4 Classification of the Finite Compounds of Cubes

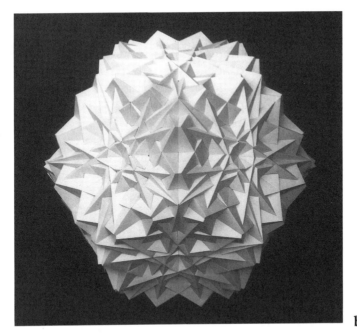

b

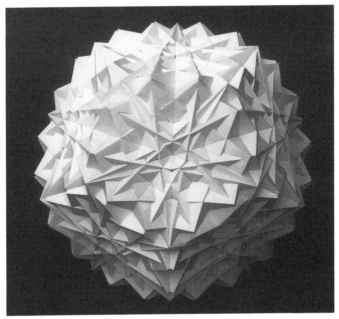

c

4.6. Diagrams of Compound Connections

Table 23 provides a summary of the compound types for each group. An abbreviated notation for the types is added to the diagrams, providing sufficient information on compound distinction for each group separately, restricted to the freedom.

The diagrams illustrate how, for each group, compounds with rotational and rigid freedom are interconnected. Compounds of the tetra- and octahedral groups are brought together in one diagram. A rigid compound has at least a dihedral stabilizer, and such a group contains more types of cyclic subgroups. Even $D_2 \times I$ contains two different "types" of subgroups $C_2 \times I$ because the index of $C_2 \times I$ in $D_2 \times I$ is 2, whereas there are three distinct such subgroups. Hence, a rigid compound represents a crossing between two or more compounds with rotational freedom. Compounds with central freedom connect all other ones. The edges, or a circle, represent compounds with rotational freedom, and the vertices compounds of rigid compounds.

Chapter 4 Classification of the Finite Compounds of Cubes

Table 23. The 30 finite compound types of cubes.

Group	Rigid	Rotational Freedom	Central Freedom	Symbol
Cyclic			$n\mid C_n \times I \mathbin{/} E \times I$	n
Dihedral	$n\mid D_{4n} \times I \mathbin{/} D_4 \times I$			n_4
	$n\mid D_{3n} \times I \mathbin{/} D_3 \times I$			n_3
	$n\mid D_{2n} \times I \mathbin{/} D_2 \times I$			n_2
		$2n\mid D_{4n} \times I \mathbin{/} C_{4n} \times I$		$2n_4$
		$2n\mid D_{3n} \times I \mathbin{/} C_{3n} \times I$		$2n_3$
		$2n\mid D_{2n} \times I \mathbin{/} D_1 \times I$		$2n_2$
		$nA\mid D_n \times I \mathbin{/} C_2 \times I$		nA
		$nB\mid D_n \times I \mathbin{/} C_2 \times I$		nB
			$2n\mid D_n \times I \mathbin{/} E \times I$	$2n$
Tetrahedral		$4\mid A_4 \times I \mathbin{/} C_3 \times I$		4
		$6\mid A_4 \times I \mathbin{/} C_2 \times I$		6
			$12\mid A_4 \times I \mathbin{/} E \times I$	12
Octahedral	$1\mid S_4 \times I \mathbin{/} S_4 \times I$			1
	$3\mid S_4 \times I \mathbin{/} D_4 \times I$			3
	$4\mid S_4 \times I \mathbin{/} D_3 \times I$			4
	$6\mid S_4 \times I \mathbin{/} D_2 \times I$			6
		$6\mid S_4 \times I \mathbin{/} C_4 \times I$		6
		$8\mid S_4 \times I \mathbin{/} C_3 \times I$		8
		$12\mid S_4 \times I \mathbin{/} D_1 \times I$		12
		$12A\mid S_4 \times I \mathbin{/} C_2 \times I$		12A
		$12B\mid S_4 \times I \mathbin{/} C_2 \times I$		12B
			$24\mid S_4 \times I \mathbin{/} E \times I$	24
Icosahedral	$5\mid A_5 \times I \mathbin{/} A_4 \times I$			5
	$10\mid A_5 \times I \mathbin{/} D_3 \times I$			10A, 10B
	$15\mid A_5 \times I \mathbin{/} D_2 \times I$			15
		$20\mid A_5 \times I \mathbin{/} C_3 \times I$		20
		$30A\mid A_5 \times I \mathbin{/} C_2 \times I$		30A
		$30B\mid A_5 \times I \mathbin{/} C_2 \times I$		30B
			$60\mid A_5 \times I \mathbin{/} E \times I$	60

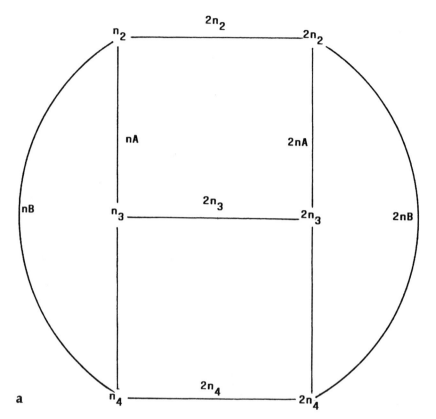

Diagram 2. (a) Standard diagram of dihedral compounds (rigid compounds may vary according to n). (b) Tetrahedral (∗) and octahedral compounds. (c) Icosahedral compounds. The circle starts and ends at 15.

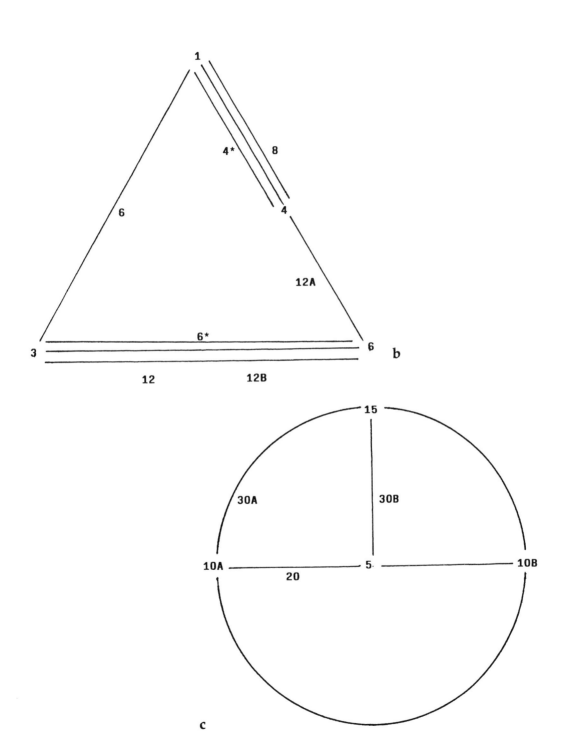

Chapter 5

Stability of Subcompounds

A next subject in the discussion of finite cube compounds concerns the relationship of a compound with its orientative subcompounds. Distinct versions contain subcompound versions that may be distinct, yet holding their number of constituents. Hence, with respect to the maximal orientative stability, the stability of a subcompound version is either basic or superior. Therefore, a *special version of the compound occurs when the basic stability of a subcompound becomes superior.*

1. Orientative Subcompounds

Let $(n_s|\mathbf{G}_s/\mathbf{F}_s)$ be a subcompound of $n|\mathbf{G}/\mathbf{H}$, in maximal orientative stability. The orientative subcompound of $n|\mathbf{G}/\mathbf{H}$ is then denoted by

$$n_s|(\mathbf{G}_s/\mathbf{F}_s) \quad \text{or} \quad n_s|(\mathbf{G}_s/\mathbf{F}_s|\mu)$$

indicating that the maximal orientative stability $(\mathbf{G}_s/\mathbf{F}_s)$ is either basic or superior. Because the number n_s of subconstituents is invariant with respect to the versatility of $n|\mathbf{G}/\mathbf{H}$, it is written outside the brackets. If the stability of all the subversions should be basic, the orientative subcompound is an actual compound $n_s|\mathbf{G}_s/\mathbf{F}_s$.

1.1. Disjunctive Subcompounds

When $\mathbf{H} \subset \mathbf{G}_s$, $n_s|(\mathbf{G}_s/\mathbf{F}_s)$ is disjunctive (see Chapter 2, Section 5.2), allowing $n|\mathbf{G}/\mathbf{H}$ to be written as a product

$$k \times n_s | (\mathbf{G}_s/\mathbf{F}_s)$$

where the number k of disjunct right coorbits is invariant with respect to the versatility of $n|\mathbf{G}/\mathbf{H}$.

If the stability of all the subversions should be basic, the notation

$$k \times n_s | \mathbf{G}_s/\mathbf{F}_s$$

expresses that the product is also that of an actual subcompound.

Let $\mathbf{H} \subset \mathbf{F} \subset \mathbf{G}$ be a proper decomposition sequence, and $n|\mathbf{G}/\mathbf{H}$ a compound. In some cases, more than one similar such decomposition line occurs, namely, when at least two distinct group representatives of the class of $\mathbf{F}$ are sharing $\mathbf{H}$. In such cases, more than one subcompound version of a same type occur, but they may be identical or congruent.

A listing of the products of disjunctive subcompounds follows next. f denotes a proper, nontrivial factor of n but is replaced by p when the factor is also prime.

1. Cyclic

1. $n | C_n \times I / E \times I$
   Prime: $p | C_p \times I / E \times I$
   Products: $n/f \times f | (C_f \times I / E \times I)$
   Hence, if $n = 2m$, the compound can be considered as $m \times 2 | (C_2 \times I / E \times I)$ (a product of $n/2$ pairs of reflected cubes), or as $2 \times m | (C_m \times I / E \times I)$, a pair of reflected compounds, $m | (C_m \times I / E \times I)$.

2. Dihedral

1. $n | D_{4n} \times I / D_4 \times I$
   Prime: $p | D_{4p} \times I / D_4 \times I$
   Products: $n/f \times f | D_{4f} \times I / D_4 \times I$
2. $n | D_{2n} \times I / D_2 \times I$
   Prime: $p | D_{2p} \times I / D_2 \times I$
   Products: $n/f \times f | D_{2f} \times I / D_2 \times I$

Chapter 5 Stability of Subcompounds

3. $n \mid D_{3n} \times I / D_3 \times I$
   Prime: $\quad p \mid D_{3p} \times I / D_3 \times I$
   Products: $n/f \times f \mid D_{3f} \times I / D_3 \times I$
4. $2n \mid D_{4n} \times I / C_4 \times I$
   Prime: $\quad 2 \mid D_4 \times I / C_4 \times I$
   Products: $2 \times n \mid D_{4n} \times I / D_4 \times I \ (n \geq 2)$
   $\quad\quad\quad\quad n \times 2 \mid (D_4 \times I / C_4 \times I)$
   $\quad\quad\quad\quad$ (prime compound of this type, $n \geq 2$)
   $\quad\quad\quad\quad 2n/f \times f \mid D_{4f} \times I / D_4 \times I$
   $\quad\quad\quad\quad n/f \times 2f \mid (D_{4f} \times I / C_4 \times I)$
   $\quad\quad\quad\quad$ (compound of this type)
5. $2n \mid D_{2n} \times I / D_1 \times I$
   Prime: $\quad 2 \mid D_2 \times I / C_2 \times I$
   Products: $2 \times n \mid D_{2n} \times I / D_2 \times I \ (n \geq 2)$
   $\quad\quad\quad\quad n \times 2 \mid (D_2 \times I / C_2 \times I)$
   $\quad\quad\quad\quad 2n/f \times f \mid D_{2f} \times I / D_2 \times I$
   $\quad\quad\quad\quad n/f \times 2f \mid (D_{2f} \times I / D_1 \times I)$
   $\quad\quad\quad\quad$ (compound of this type)
6. $2n \mid D_{3n} \times I / C_3 \times I$
   Prime: $\quad 2 \mid D_6 \times I / C_3 \times I$
   Products: $2 \times n \mid D_{3n} \times I / D_3 \times I \ (n \geq 2)$
   $\quad\quad\quad\quad n \times 2 \mid D_6 \times I / D_3 \times I$
   $\quad\quad\quad\quad$ (prime compound of this type, $n \geq 2$)
   $\quad\quad\quad\quad 2n/f \times f \mid D_{3f} \times I / D_3 \times I$
   $\quad\quad\quad\quad n/f \times 2f \mid (D_{3f} \times I / C_3 \times I)$
7. $nA \mid D_n \times I / C_2 \times I$
   Prime: $\quad pA \mid D_p \times I / C_2 \times I$
   Products: $n/f \times fA \mid (D_f \times I / C_2 \times I)$
   $\quad\quad\quad\quad$ (compound of this type)
   $\quad\quad\quad\quad \diamond \ n = 2m: m \times 2 \mid (D_2 \times I / C_2 \times I)$
8. $nB \mid D_n \times I / C_2 \times I$
   Prime: $\quad pB \mid D_p \times I / C_2 \times I$
   Products: $n/f \times fB \mid (D_f \times I / C_2 \times I)$
   $\quad\quad\quad\quad$ (compound of this type)
   $\quad\quad\quad\quad \diamond \ n = 2m: m \times 2 \mid (D_4 \times I / C_2 \times I)$

9. $2n \mid D_n \times I \,/\, E \times I$
Products: $2 \times n \mid (C_n \times I \,/\, E \times I)$
$n \times 2 \mid (C_2 \times I \,/\, E \times I)$
$n/f \times 2f \mid (D_f \times I \,/\, E \times I)$
$2n/f \times f \mid (C_f \times I \,/\, E \times I)$

3. Tetrahedral

1. $4 \mid A_4 \times I \,/\, C_3 \times I$
   Prime compound
2. $6 \mid A_4 \times I \,/\, C_2 \times I$
   Product: $3 \times 2 \mid (D_2 \times I \,/\, C_2 \times I)$
3. $12 \mid A_4 \times I \,/\, E \times I$
   Products: $3 \times 4 \mid (D_2 \times I \,/\, E \times I)$
   $4 \times 3 \mid (C_3 \times I \,/\, E \times I)$
   $6 \times 2 \mid (C_2 \times I \,/\, E \times I)$

4. Octahedral

1. $1 \mid S_4 \times I \,/\, S_4 \times I$
   Prime compound
2. $3 \mid S_4 \times I \,/\, D_4 \times I$
   Prime compound
3. $4 \mid S_4 \times I \,/\, D_3 \times I$
   Prime compound
4. $6 \mid S_4 \times I \,/\, D_2 \times I$
   Products: $3 \times 2 \mid D_4 \times I \,/\, D_2 \times I$
5. $6 \mid S_4 \times I \,/\, C_4 \times I$
   Products: $3 \times 2 \mid (D_4 \times I \,/\, C_4 \times I)$
6. $8 \mid S_4 \times I \,/\, C_3 \times I$
   Products: $2 \times 4 \mid (A_4 \times I \,/\, C_3 \times I)$
   $4 \times 2 \mid (D_3 \times I \,/\, C_3 \times I)$
7. $12 \mid S_4 \times I \,/\, D_1 \times I$
   Products: $2 \times 6 \mid (A_4 \times I \,/\, C_2 \times I)$
   $3 \times 4 \mid (D_4 \times I \,/\, D_1 \times I)$
   $6 \times 2 \mid (D_2 \times I \,/\, C_2 \times I)$

Chapter 5 Stability of Subcompounds   165

    8. $12A \mid S_4 \times I \mid C_2 \times I$
       Products: $3 \times 4A \mid (D_4 \times I \mid C_2 \times I)$
                 $4 \times 3A \mid (D_3 \times I \mid C_2 \times I)$
                 $6 \times\ \ 2 \mid (D_2 \times I \mid C_2 \times I)$
    9. $12B \mid S_4 \times I \mid C_2 \times I$
       Products: $3 \times 4B \mid (D_4 \times I \mid C_2 \times I)$
                 $4 \times 3B \mid (D_3 \times I \mid C_2 \times I)$
                 $6 \times\ \ 2 \mid (D_4 \times I \mid C_4 \times I)$
  10. $24 \mid S_4 \times I \mid E \times I$
       Products: $\ 2 \times 12 \mid (A_4 \times I \mid E \times I)$
                $\ 3 \times\ \ 8 \mid (D_4 \times I \mid E \times I)$
                $\ 4 \times\ \ 6 \mid (D_3 \times I \mid E \times I)$
                $\ 6 \times\ \ 4 \mid (C_4 \times I \mid E \times I)$
                $\ 6 \times\ \ 4 \mid (D_2 \times I \mid E \times I)$
                $\ 8 \times\ \ 3 \mid (C_3 \times I \mid E \times I)$
                $12 \times\ \ 2 \mid (D_1 \times I \mid E \times I)$
                $12 \times\ \ 2 \mid (C_2 \times I \mid E \times I)$

5. *Icosahedral*

    1. $5 \mid A_5 \times I \mid A_4 \times I$
       Prime compound
    2. $10 \mid A_5 \times I \mid D_3 \times I$
       Prime compound
    3. $15 \mid A_5 \times I \mid D_2 \times I$
       Products: $5 \times 3 \mid S_4 \times I \mid D_4 \times I$
    4. $20 \mid A_5 \times I \mid C_3 \times I$
       Products: $5 \times 4 \mid (A_4 \times I \mid C_3 \times I)$
    5. $30A \mid A_5 \times I \mid C_2 \times I$
       Products: $\ \ 5 \times\ \ 6 \mid (A_4 \times I \mid C_2 \times I)$
                $15 \times\ \ 2 \mid (D_2 \times I \mid C_2 \times I)$
                $\ \ 6 \times 5A \mid (D_5 \times I \mid C_2 \times I)$
                $10 \times 3A \mid (D_3 \times I \mid C_2 \times I)$
    6. $30B \mid A_5 \times I \mid C_2 \times I$
       Products: $\ \ 5 \times\ \ 6 \mid S_4 \times I \mid C_4 \times I$
                $15 \times\ \ 2 \mid (D_4 \times I \mid C_4 \times I)$
                $\ \ 6 \times 5B \mid (A_5 \times I \mid C_2 \times I)$
                $10 \times 3B \mid (A_5 \times I \mid C_2 \times I)$

7. $60 | A_5 \times I / E \times I$
Products: $5 \times 12 | (A_4 \times I / E \times I)$
$\phantom{\text{Products: }}10 \times 6 | (D_3 \times I / E \times I)$
$\phantom{\text{Products: }}15 \times 4 | (D_2 \times I / E \times I)$
$\phantom{\text{Products: }}20 \times 3 | (C_3 \times I / E \times I)$
$\phantom{\text{Products: }}30 \times 2 | (C_2 \times I / E \times I)$

1.2. Conjunctive Subcompounds

When $\mathbf{H} \not\subset \mathbf{G}_s$, $n_s | (\mathbf{G}_s / \mathbf{F}_s)$ is conjunctive (see Chapter 2, Section 5.1), an upgrade from basic to superior stability provides extra information on the coincidence of *constitutional axes*, i.e., the symmetry axes of the constituents. This type of subcompound occurs only with rotational and rigid freedom because $E \times I \subset \mathbf{H}$ must be proper. A conjunctive cyclic subcompound $m | (C_m \times I / E \times I)$ then acquires a superior stability when an axis of the descriptive is perpendicular to, or coincidental with, the sole axis of $C_m \times I$ (see Chapter 4, Section 3.3, item 1). As the superior stability is at least dihedral, m constitutional axes become coincidental. The number of such coincidental sets is then the index of $C_m \times I$ in $\mathbf{G}$, which is $g/2m$.

$n | \mathbf{G}/\mathbf{H}$ has the following theoretical number of constitutional axes:

twofold: $6g/h$
threefold: $4g/h$
fourfold: $3g/h$

Let $\mathbf{G}$ be higher than dihedral, $\mathbf{H} = C_2 \times I$, and a descriptive coincidence $2 \mapsto 2$ occurs. Then, the number of shared pairs of constitutional twofold axes is $g/4$, and number of remaining constitutional axes is

twofold: $3g/2$
threefold: g
fourfold: $3g/4$

An axial coincidence with a mirror (which is then always the sole mirror of a conjunctive subgroup $C_2 \times I$) results in $g/4$ pairs of constitutional axes. Hence, in neither of the cases does the totality of remaining constitutional axes of an equal type form coinciding pairs. If $\mathbf{H} = C_d \times I$ ($d = 3, 4$), the number of remaining twofold axes is even larger. Hence, the coincidence of twofold constitutional axes will not be considered for special compound versions.

2. Discussion of Rigid Freedom

The only rigid compound with a fundamental versatility of more than one described position is $10 |\ A_5 \times I\ /\ D_3 \times I$. This prime compound with two versions is very particular, which will become clear by the discussion of its conjunctive subcompounds with rotational freedom.

2.1. Subcompound $2|\ (D_2 \times I\ /\ C_2 \times I)$

    a. Version A
        $2|\ D_2 \times I\ /\ C_2 \times I\ |14°21'33''.24$ (Fig. 114)
    b. Version B
        $2|\ (D_2 \times I\ /\ C_2 \times I\ |56°10'10''.36)$
        $= 2|\ D_2 \times I\ /\ C_2 \times I\ |33°49'49''.64$

2.2. Subcompound $6|\ (A_4 \times I\ /\ C_2 \times I)$

    a. Version A
        $2|\ A_4 \times I\ /\ C_2 \times I\ |14°21'33''.24$
    b. Version B
        $(2|\ A_4 \times I\ /\ C_2 \times I\ |56°10'10''.36)$
        $= 2|\ A_4 \times I\ /\ C_2 \times I\ |33°49'49''.64$

2.3. Subcompound $2|\ (D_3 \times I\ /\ C_3 \times I)$

    a. Version A
        $2|\ (D_3 \times I\ /\ C_3 \times I\ |0°) = 1|\ S_4 \times I\ /\ S_4 \times I$
    b. Version B
        $2|\ (D_3 \times I\ /\ C_3 \times I\ |60°) = 1|\ S_4 \times I\ /\ S_4 \times I$

Figure 114. View of $6 \mid A_4 \times I \mid C_2 \times I \mid 14°21'33''.24$, along a twofold axis

2.4. Subcompound $4 \mid (A_4 \times I \mid C_3 \times I)$

The icosahedral compound $(20 \mid A_5 \times I \mid C_3 \times I)$ contains two different versions of the orientative subcompound $(4 \mid A_4 \times I \mid C_3 \times I)$ simultaneously. The rotational freedom of $(20 \mid A_5 \times I \mid C_3 \times I)$ can then be observed from two different initial reference positions (see Fig. 87), for which

$$\mu = 44°14'39''.04$$

The distinct versions of the tetrahedral subcompound are

$$(4 \mid A_4 \times I \mid C_3 \times I \mid \mu_a)$$
$$(4 \mid A_4 \times I \mid C_3 \times I \mid \mu_b)$$

where

$$\mu_a = \mu_b + 44°28'39''.04$$

In the opposite case, any version $(4 \mid A_4 \times I \mid C_3 \times I \mid \mu)$ determines

Chapter 5 Stability of Subcompounds

Figure 115. View of $4 \mid A_4 \times I / C_3 \times I \mid 37°45'40''.48$, along a three-fold axis

two distinct versions of $(20 \mid A_5 \times I / C_3 \times I)$ that each contain $(4 \mid A_4 \times I / C_3 \times I \mid \mu)$ as a subcompound:

$$(20 \mid A_5 \times I / C_3 \times I \mid \mu)$$
$$(20 \mid A_5 \times I / C_3 \times I \mid -\mu)$$

a. Version A

$10 \mid A_5 \times I / D_3 \times I \mid A = (20 \mid A_5 \times I / C_3 \times I \mid 37°45'40''.48)$

The rotational angles of the two conjunctive subcompounds $4 \mid (A_4 \times I / C_3 \times I \mid \mu)$, whose stabilizer $C_3 \times I \not\supset \mathbf{H}$ ($= D_3 \times I$), are

$\mu_a = 37°45'40''.48$
$\mu_b = 37°45'40''.48 + 44°28'39''.04 = 82°14'19''.52$

Hence, both conjunctive subcompound versions are identical (Fig. 115), being

$$4 \mid A_4 \times I / C_3 \times I \mid 37°45'40''.48$$

Moreover, when the subcompound $4|\ A_4 \times I\ /\ C_3 \times I\ |37°45'40".48$ is deleted in $10|\ A_5 \times I\ /\ D_3 \times I\ |A$, the set of the six remaining cubes is (Fig. 114)

$$6|\ A_4 \times I\ /\ C_2 \times I\ |14°21'33".24$$

This is easily understood by applying an "alternative" argumentation. Because a compound $10|\ A_5 \times I\ /\ D_3 \times I$ is invariant under the subgroup $A_4 \times I \supset \mathbf{H}$, and $4|\ A_4 \times I\ /\ C_3 \times I$ is invariant under this subgroup (it is its own subaction group), the remaining set of six cubes is invariant under it too. Hence, this set is a version of $6|\ (A_4 \times I\ /\ C_2 \times I\ |\mu)$. The angle of rotation is identical to that of $30A|\ A_5 \times I\ /\ C_2 \times I\ |\mu$.

Hence, the compound is the following sum of tetrahedral compounds (Fig. 116):

$$10|\ A_5 \times I\ /\ D_3 \times I\ |A\ =\ 4|\ A_4 \times I\ /\ C_3 \times I\ |37°45'40".48$$
$$+\ 6|\ A_4 \times I\ /\ C_2 \times I\ |14°21'33".24$$

b. Version B

$$10|\ A_5 \times I\ /\ D_3 \times I\ |B\ =\ (20|\ A_5 \times I\ /\ C_3 \times I\ |-22°14'19".52)$$

The rotational angles of the two conjunctive subcompounds $4|\ (A_4 \times I\ /\ C_3 \times I\ |\mu)$, whose stabilizer $C_3 \times I \not\supset \mathbf{H}$ $(= D_3 \times I)$, are

$$\mu_a\ =\ -22°14'19".52$$
$$\mu_b\ =\ -22°14'19".52 + 44°28'39".04\ =\ 22°14'19".52$$

Hence, both conjunctive subcompound versions are identical (Fig. 117), being

$$4|\ A_4 \times I\ /\ C_3 \times I\ |22°14'19".52$$

The remark in Version A holds also for this version: After subtraction of the subcompound $4|\ A_4 \times I\ /\ C_3 \times I$ $|22°14'19".52$, the tetrahedral subcompound

$$6|\ A_4 \times I\ /\ C_2 \times I\ |33°49'49".64$$

remains. Hence,

Chapter 5 Stability of Subcompounds

Figure 116. View of 10 | $A_5 \times I$ / $D_3 \times I$ | A, along a threefold axis, as the sum of the subcompounds in Figs. 114 and 115.

Figure 117. View of 4 | $A_4 \times I$ / $C_3 \times I$ | 22°14′19″.52, along a threefold axis

$$10|\,A_5 \times I\,/\,D_3 \times I\,|B\ =\ 4|\,A_4 \times I\,/\,C_3 \times I\,|22°14'19''.52$$
$$+\ 6|\,A_4 \times I\,/\,C_2 \times I\,|33°49'49''.64$$

Moreover, when the sum of the two versions of $10|\,A_5 \times I\,/\,D_3 \times I$ is taken, an arrangement of 20 cubes is obtained, whose symmetry group is $A_5 \times I$ but is not a compound by the definition (see also Chapter 6, Section 3.1). However, this arrangement is peculiar in the sense that a dihedral compound $2|\,D_6 \times I\,/\,D_3 \times I$ occurs about each of the 10 threefold axes: a sixfold axis of $2|\,D_6 \times I\,/\,D_3 \times I$ is coincidental with a threefold axis of $A_5 \times I$. Hence,

$$10|\,A_5 \times I\,/\,D_3 \times I\,|A\ +\ 10|\,A_5 \times I\,/\,D_3 \times I\,|B$$
$$=\ 10\ \times\ 2|\,D_6 \times I\,/\,D_3 \times I$$

3. Discussion of Rotational Freedom

3.1. Tetrahedral

1. $4|\,A_4 \times I\,/\,C_3 \times I$

In Fig. 118 we show how a special version occurs for $\mu_2 = 44°28'39''.04$ when a conjunctive subcompound $2|\,(C_2 \times I\,/\,E \times I)$ obtains a superior stability $D_6 \times I\,/\,D_3 \times I$ because a threefold axis is coplanar with the mirror of a subgroup $C_2 \times I \not\supset \mathbf{H}$.

In addition to the four threefold axes of $A_4 \times I$, the compound has 12 more constitutional threefold axes, totaling 16, and has 32 vertices (2 on each axis). The conjunctive subcompound $2|\,(C_2 \times I\,/\,E \times I)$ in this special version is $2|\,D_3 \times I\,/\,C_3 \times I\,|22°14'19''.52$ (Fig. 119). As such, a pair of threefold axes occurs for each of the 6 right coorbits, altogether consisting of the 12 remaining threefold axes. *Hence, all constitutional threefold axes, apart from those of $A_4 \times I$, form coincidental pairs.* Moreover, the compound version has 20 vertices (Fig. 120).

Remarkably, $4|\,A_4 \times I\,/\,C_3 \times I\,|44°28'39''.04$ is a sum of two distinct compounds $2|\,D_3 \times I\,/\,C_3 \times I\,|22°14'19''.52$, which can be taken in three different combinations of the six occurring right coorbits (Fig. 121).

To understand the more specific properties of this special version, it must be embedded in the higher group $A_5 \times I$. The

Chapter 5   Stability of Subcompounds

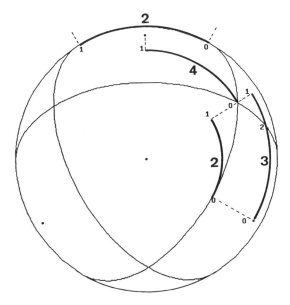

Figure 118. Special positions of $4\,|\,A_4 \times I\,/\,C_3 \times I$ in Fig. 85

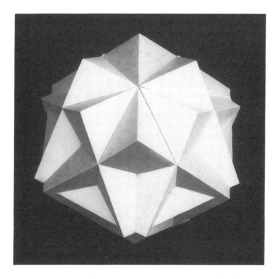

Figure 119. View of $2\,|\,D_3 \times I\,/\,C_3 \times I\,|\,22°14'19''.52$, along a threefold axis

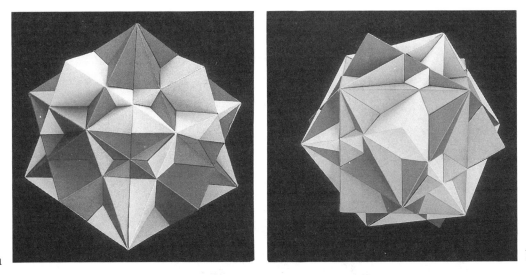

Figure 120. $4\,|\,A_4\times I\,/\,C_3\times I\,|\,44°28'39''.04$; (a) along a twofold axis; (b) along a threefold axis

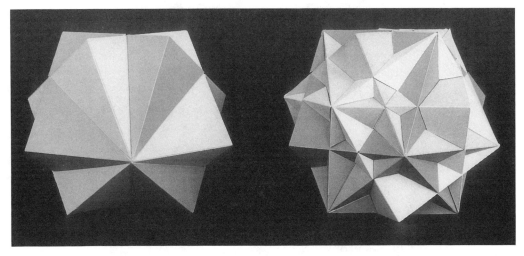

Figure 121. $2\,|\,D_3\times I\,/\,C_3\times I\,|\,22°14'19''.52$ in $4\,|\,A_4\times I\,/\,C_3\times I\,|\,44°28'39''.04$

two versions of the icosahedral compound $(20|\,A_5\times I\,/\,C_3\times I)$, in each of which a tetrahedral disjunctive subcompound $4|\,A_4\times I\,/\,C_3\times I\,|\mu_2$ is contained, are (see Section 2.4)

(1) $(20|\,A_5\times I\,/\,C_3\times I\,|44°28'39''.04)$
$= 20|\,A_5\times I\,/\,C_3\times I\,|31°02'41''.91$

(2) $(20|\,A_5\times I\,/\,C_3\times I\,|-44°28'39''.04)$
$= (20|\,A_5\times I\,/\,C_3\times I\,|0°) = 5|\,A_5\times I\,/\,A_4\times I$

Hence, from (2) follows that *any four cubes out of the classic compound of five cubes constitute the special version* $4|\,A_4\times I\,/\,C_3\times I\,|\mu_2$. This peculiar property can also be derived from an alternative argumentation, similar to the one in Section 2.4a. The compound of five cubes is invariant under any *I*-subgroup of $A_5\times I$, among which is the stabilizer $A_4\times I$. Because the descriptive is invariant under the stabilizer, so is the remaining set of four cubes. Hence, any four cubes in the classic compound of five cubes is $4|\,A_4\times I\,/\,C_3\times I\,|44°28'39''.04$. As a consequence, *the casing of* $4|\,A_4\times I\,/\,C_3\times I\,|44°28'39''.04$ *is a regular dodecahedron.*

2. $6|\,A_4\times I\,/\,C_2\times I$

A special position is found in Fig. 122.

$6|\,A_4\times I\,/\,C_2\times I = 3 \times 2|\,D_2\times I\,/\,C_2\times I$, except for $\mu_2 = 35°15'51''.80$ where the disjunctive subcompound version $2|\,(D_2\times I\,/\,C_2\times I)$ has a superior stability $D_6\times I\,/\,D_3\times I$ and becomes rigid. (See Chapter 4, Section 4.2.6, $n = 1$; corresponding to the intermediate position.) Hence,

$6|\,A_4\times I\,/\,C_2\times I\,|35°15'51''.80 = 3 \times 2|\,D_6\times I\,/\,D_3\times I$

3.2. Octahedral

1. $6|\,S_4\times I\,/\,C_4\times I$

A special position is found in Fig. 123.

$6|\,S_4\times I\,/\,C_4\times I = 3 \times 2|\,D_4\times I\,/\,C_4\times I$, except for $\mu_2 = 22°30'$ where the disjunctive subcompound version $2|\,(D_4\times I\,/\,C_4\times I)$ has a superior stability $D_8\times I\,/\,D_4\times I$ and becomes rigid

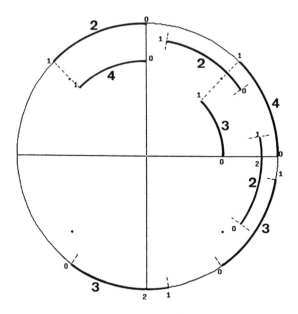

Figure 122. Special positions of $6 \mid A_4 \times I \mathbin{/} C_2 \times I$ in Fig. 88

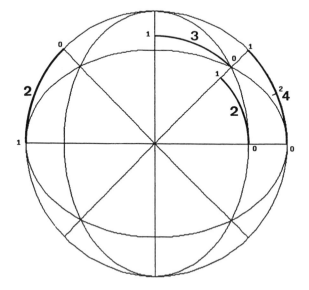

Figure 123. Special positions of $6 \mid S_4 \times I \mathbin{/} C_4 \times I$ in Fig. 84

Chapter 5 Stability of Subcompounds

(see Chapter 4, Section 4.2.4, $n = 1$; corresponding to the end position). Hence,

$$6|\, S_4 \times I\, /\, C_4 \times I\, |22°30' = 3 \times 2|\, D_8 \times I\, /\, D_4 \times I$$

2. $8|\, S_4 \times I\, /\, C_3 \times I$

Two special positions are found in Fig. 124.

a. For $\mu_2 = 44°28'39''.04$, a conjunctive subcompound $2|\,(C_2 \times I\, /\, E \times I)$ obtains a superior stability $D_6 \times I\, /\, D_3 \times I$ due to the coplanarity of a threefold axis with a mirror of a subgroup $C_2 \times I \not\supset \mathbf{H}$.

Because the compound version is composed of four pairs of cubes, each sharing a threefold axis with $S_4 \times I$, the number of remaining constitutional threefold axes is $32 - 8 = 24$. $2|\,(C_2 \times I\, /\, E \times I)$ appears here as the same version of Section 3.1.1, namely, $2|\, D_3 \times I\, /\, C_3 \times I$

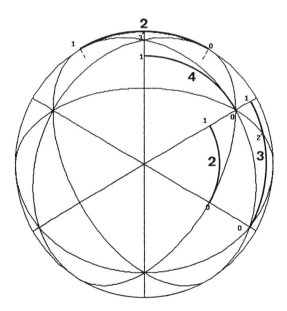

Figure 124. Special positions of $8|\, S_4 \times I\, /\, C_3 \times I$ in Fig. 86

Figure 125. View of 8| $S_4 \times I$ / $C_3 \times I$ |44°28′39″.04, along a threefold axis

|22°14′19″.52. In the 12 right coorbits of 2| ($C_2 \times I$ / $E \times I$), 12 pairs of threefold axes occur, altogether consisting of the 24 remaining axes. Hence, *all constitutional threefold axes form coincidental pairs.* Moreover, the compound has 32 vertices, and each is shared by a pair of cubes. Eight (2 × 4) double vertices lie on the threefold axes of $S_4 \times I$, and 24 (2 × 12) double vertices are lying outside these axes in mirrors of the group. Hence, 8| $S_4 \times I$ / $C_3 \times I$ |44°28′39″.04 has *two constituent cubes per vertex* (Fig. 125). This version was also illustrated in Chapter 4, Section 4.4.6.

The eight single vertices of the disjunctive subcompound 4| $A_4 \times I$ / $C_3 \times I$ |44°28′39″.04 are those of a cube, which shall be called (**C**). This cube shares a fourfold axis with $S_4 \times I$ (being a twofold axis of $A_4 \times I$) but is *not* a constituent of the compound. The product is

$$8|\ S_4 \times I\ /\ C_3 \times I\ |44°28'39''.04$$
$$= 2 \times 4|\ A_4 \times I\ /\ C_3 \times I\ |44°28'39''.04$$

The two compounds $4|\ A_4 \times I\ /\ C_3 \times I\ |44°28'39''.04$ are rotated images of one another through 90° about a twofold axis of $A_4 \times I$. When a tetrahedral compound is rotated about that axis through 90°, the cube (**C**) is coinciding with itself. Each of its vertices is then becoming a vertex shared by a pair of cubes in $8|\ S_4 \times I\ /\ C_3 \times I\ |44°28'39''.04$, and these are not coincidental because successive cubes in $4|\ A_4 \times I\ /\ C_3 \times I$ have rotated in opposite senses (see Chapter 4, Section 4.3.1).

In all positions $8|\ S_4 \times I\ /\ C_3 \times I\ |\mu$, the tetrahedral subcompound is $4|\ A_4 \times I\ /\ C_3 \times I\ |\mu$. *The casing of* $8|\ S_4 \times I\ /\ C_3 \times I\ |44°28'39''.04$ *is a compound of two dodecahedra* whose symmetry group is $S_4 \times I$. The two dodecahedra share a twofold axis, which is coincidental with a fourfold axis of $S_4 \times I$ and are images of one another under a rotation through 90° about that axis.

b. $8|\ S_4 \times I\ /\ C_3 \times I = 4 \times 2|\ D_3 \times I\ /\ C_3 \times I$, except for $\mu_3 = 30°$, where the disjunctive subcompound version $2|\ (D_3 \times I\ /\ C_3 \times I)$ has a superior stability $D_6 \times I\ /\ D_3 \times I$ and becomes rigid (see Chapter 4, Section 4.2.5, $n = 1$; corresponding to the end position). Hence,

$$8|\ S_4 \times I\ /\ C_3 \times I\ |30° = 4 \times 2|\ D_6 \times I\ /\ D_3 \times I$$

3. $12|\ S_4 \times I\ /\ D_1 \times I$

Three special positions are found in Fig. 126.

a. $12|\ S_4 \times I\ /\ D_1 \times I = 3 \times 4A|\ D_4 \times I\ /\ C_2 \times I$, except for $\mu_3 = 35°15'51''.80$ where the disjunctive subcompound version $4A|\ (D_4 \times I\ /\ C_2 \times I)$ has a superior stability and becomes rigid (see Chapter 4, Section 4.2.7, $n = 4$; corresponding to the position of additional coincidence $3 \mapsto 4$, where 3 is coprime with 4). Hence,

$$12|\ S_4 \times I\ /\ D_1 \times I\ |35°15'51''.80 = 3 \times 4|\ D_{12} \times I\ /\ D_3 \times I$$

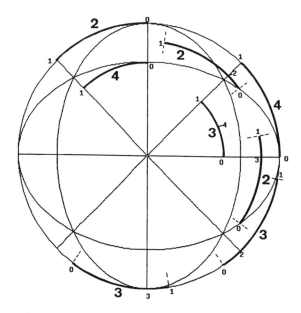

Figure 126. Special positions of $12|\ S_4 \times I\ /\ D_1 \times I$ in Fig. 89

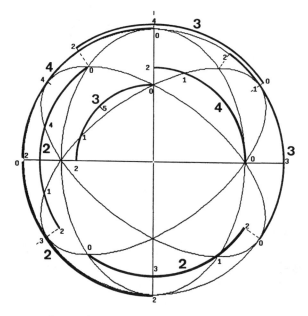

Figure 127. Special positions of $12A|\ S_4 \times I\ /\ C_2 \times I$ in Fig. 90

b. $12|\,S_4 \times I\,/\,D_1 \times I = 6 \times 2|\,D_2 \times I\,/\,C_2 \times I\,(2\times)$, except for $\mu_2 = 9°44'08''.20$ and $\mu_3 = 35°15'51''.80$ where alternatively one of both disjunctive subcompound versions $2|\,(D_2 \times I\,/\,C_2 \times I)$ has a superior stability $D_6 \times I\,/\,D_3 \times I$ and becomes rigid (see Chapter 4, Section 4.2.6, $n = 1$; corresponding to the intermediate position). Hence,

$$6 \times 2|\,D_6 \times I\,/\,D_3 \times I = 12|\,S_4 \times I\,/\,D_1 \times I\,|\,9°44'08''.20$$
$$= 12|\,S_4 \times I\,/\,D_1 \times I\,|\,35°15'51''.80$$

c. $12|\,S_4 \times I\,/\,D_1 \times I = 3 \times 4|\,D_4 \times I\,/\,D_1 \times I$, except for $\mu_4 = 22°30'$ where the disjunctive subcompound version $4|\,(D_4 \times I\,/\,D_1 \times I)$ has a superior stability $D_8 \times I\,/\,D_2 \times I$ and becomes rigid (see Chapter 4, Section 4.2.6, $n = 2$; corresponding to the end position). Hence,

$$12|\,S_4 \times I\,/\,D_1 \times I\,|\,22°30' = 3 \times 4|\,D_8 \times I\,/\,D_2 \times I$$

In conclusion,

$$12|\,S_4 \times I\,/\,D_1 \times I\,|\,9°44'08''.20 = 6 \times 2|\,D_6 \times I\,/\,D_3 \times I$$
$$12|\,S_4 \times I\,/\,D_1 \times I\,|\,22°30' = 3 \times 4|\,D_8 \times I\,/\,D_2 \times I$$
$$12|\,S_4 \times I\,/\,D_1 \times I\,|\,35°15'51''.80 = 3 \times 4|\,D_{12} \times I\,/\,D_3 \times I$$
$$= 6 \times 2|\,D_6 \times I\,/\,D_3 \times I$$

4. $12A|\,S_4 \times I\,/\,C_2 \times I$

   Three special positions are found in Fig. 127.

   a. $12A|\,S_4 \times I\,/\,C_2 \times I = 3 \times 4A|\,D_4 \times I\,/\,C_2 \times I$, except for $\mu_4 = 54°44'08''.20$ where the disjunctive subcompound version $4A|\,(D_4 \times I\,/\,C_2 \times I)$ has a superior stability $D_{12} \times I\,/\,D_3 \times I$ and becomes rigid (see Chapter 4, Section 4.2.7, $n = 4$; corresponding to the position of additional coincidence $3 \mapsto 4$, where 4 is coprime with 3). Hence,

   $$12A|\,S_4 \times I\,/\,C_2 \times I\,|\,54°44'08''.20 = 3 \times 4|\,D_{12} \times I\,/\,D_3 \times I$$

   b. $12A|\,S_4 \times I\,/\,C_2 \times I = 4 \times 3A|\,D_3 \times I\,/\,C_2 \times I\,(2\times)$, except for $\mu_3 = 35°15'51''.80$ and $\mu_4 = 54°44'08''.20$ where alternatively one of the two subcompound versions $3A|\,(D_3 \times I\,/\,C_2 \times I)$ has a superior stability, respectively,

$D_6 \times I / D_2 \times I$ and $D_{12} \times I / D_4 \times I$ and becomes rigid (see Chapter 4, Section 4.2.7, $n = 3$; respectively, corresponding to the end position and the initial position). Hence,

$12A | S_4 \times I / C_2 \times I | 35°15'51".80 = 4 \times 3 | D_6 \times I / D_2 \times I$
$12A | S_4 \times I / C_2 \times I | 54°44'08".20 = 4 \times 3 | D_{12} \times I / D_4 \times I$

c. $12A | S_4 \times I / C_2 \times I = 6 \times 2 | D_2 \times I / C_2 \times I$, except for μ_3 and μ_4 where the disjunctive subcompound versions $2 | (D_2 \times I / C_2 \times I)$ have a superior stability $D_6 \times I / D_3 \times I$ (see Chapter 4, Section 4.2.6, $n = 1$; both corresponding to the version $(2 | D_2 \times I / C_2 \times I | 35°15'51".80)$, and for $\mu_5 = 45°$ where it has a superior stability $D_4 \times I / D_2 \times I$ [corresponding to the end position of $(2 | D_2 \times I / C_2 \times I)$]. In each of these versions, it becomes rigid. Hence,

$6 \times 2 | D_6 \times I / D_3 \times I = 12A | S_4 \times I / C_2 \times I | 35°15'51".80$
$= 12A | S_4 \times I / C_2 \times I | 54°44'08".20$
$6 \times 2 | D_4 \times I / D_2 \times I = 12A | S_4 \times I / C_2 \times I | 45°$

In conclusion,

$12A | S_4 \times I / C_2 \times I | 35°15'51".80 = 4 \times 3 | D_6 \times I / D_2 \times I$
$= 6 \times 2 | D_6 \times I / D_3 \times I$
$12A | S_4 \times I / C_2 \times I | 54°44'08".20 = 4 \times 3 | D_{12} \times I / D_4 \times I$
$= 6 \times 2 | D_6 \times I / D_3 \times I$
$12A | S_4 \times I / C_2 \times I | 45° = 6 \times 2 | D_4 \times I / D_2 \times I$

5. $12B | S_4 \times I / C_2 \times I$

Three special positions are found in Fig. 128.

a. $12B | S_4 \times I / C_2 \times I = 4 \times 3B | D_3 \times I / C_2 \times I (2\times)$, except for $\mu_2 = 9°44'08".20$ and $\mu_3 = 35°15'51".80$, where alternatively one of both disjunctive subcompounds $3B | (D_3 \times I / C_2 \times I)$ has a superior stability, respectively,

Chapter 5 Stability of Subcompounds

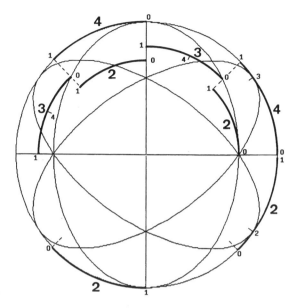

Figure 128. Special positions of $12B|S_4\times I / C_2\times I$ in Fig. 91

$D_6\times I / D_2\times I$ and $D_{12}\times I / D_4\times I$, and becomes rigid (see Chapter 4, Section 4.2.8, $n = 3$; respectively corresponding to the end and the initial positions). Hence,

$12B|S_4\times I / C_2\times I |9°44'08".20 = 4 \times 3| D_6\times I / D_2\times I$
$12B|S_4\times I / C_2\times I |35°15'51".80 = 4 \times 3| D_{12}\times I / D_4\times I$

b. $12B|S_4\times I / C_2\times I = 6 \times 2| D_4\times I / C_4\times I$, except for $\mu_4 = 22°30'$ where the disjunctive subcompound $2|(D_4\times I / C_4\times I)$ has a superior stability $D_8\times I / D_4\times I$ and becomes rigid (see Chapter 4, Section 4.2.4, $n = 1$; corresponding to the end position). Hence,

$12B|S_4\times I / C_2\times I |22°30' = 6 \times 2| D_8\times I / D_4\times I$

3.3. Icosahedral

1. $20|A_5\times I / C_3\times I$

Four special positions are found in Fig. 129.

a. For $\mu_3 = 31°02'41''.91$, a conjunctive subcompound $2|\,(C_2 \times I \,/\, E \times I)$ obtains a superior stability $D_3 \times I \,/\, C_3 \times I$ because a threefold axis is coplanar with the mirror of a subgroup $C_2 \times I \not\supset \mathbf{H}$.

Each of the 10 threefold axes of $A_4 \times I$ is shared by a pair of cubes in this compound. The number of the remaining constitutional threefold axes is then $80 - 20 = 60$. Also here, the conjunctive subcompound version is $2|\, D_3 \times I \,/\, C_3 \times I \,|22°14'19''.52$, as in Section 3.1.1 and 3.2.2a. In each of the 30 right coorbits, a threefold axis is shared by a pair of cubes; the total constitutes the remaining 60 axes. Hence, *all constitutional threefold axes are shared by pairs of cubes*. In this compound version, there are $2 \times (10 + 30) = 80$ vertices, all shared by a pair of cubes.

Hence, *the special version has two constituents per vertex*. Twenty of these lay on the threefold axes of $A_5 \times I$ and form the vertices of a dodecahedron, which shall be denoted by (**D**), whereas 60 lay outside these axes in mirrors.

The model illustrated in Figs. 130a–130c is actually a "simplified" version. A number of extremely small additional parts are deleted, which are shown in a closeup along the fivefold axis in Fig. 130d. The rotational angles of the two subcompounds $(4|\, A_4 \times I \,/\, C_3 \times I \,|\mu)$ are

$$\mu_a = 31°02'41''.91$$
$$\mu_b = 31°02'41''.91 + 44°28'39''.04 = 75°31'20''.95$$

Hence,

$$20|\, A_5 \times I \,/\, C_3 \times I \,|31°02'41''.91$$
$$= 5 \times 4|\, A_4 \times I \,/\, C_3 \times I \,|31°02'41''.91$$
$$= 5 \times 4|\, A_4 \times I \,/\, C_3 \times I \,|44°28'39''.04$$

The second subcompound is also the special version as described in Section 3.1.1.

Chapter 5 Stability of Subcompounds

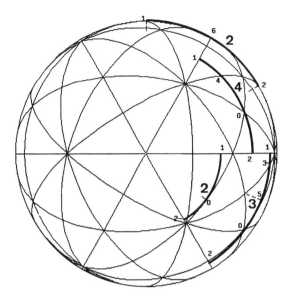

Figure 129. Special positions of $20 \mid A_5 \times I \mid C_3 \times I$ in Fig. 87

The casing of this version is a compound of five dodecahedra under icosahedral symmetry. The dodecahedron (**D**) is *not* a constituent of this latter compound, as may be easily observed in Fig. 130, but it is congruent with the constituents of it. That compound is obtained as follows: Take the five edges of one pentagonal face of (**D**) and rotate (**D**) through 90° about each of the five twofold axes, which pass through the mid-edges.

b. For $\mu_4 = 23°25'51''.28$, a conjunctive subcompound $2 \mid (C_2 \times I \mid E \times I)$ obtains a superior stability $D_4 \times I \mid C_4 \times I$ because a fourfold axis is coplanar with the mirror of a subgroup $C_2 \times I \not\supset \mathbf{H}$.

There are 60 constitutional fourfold axes in this compound and 120 constitutional squares. Then, the 30 right coorbits have one such axis shared by a pair of cubes. Hence, *all constitutional fourfold axes are shared by a pair of cubes*. Therefore, the compound version is composed of 60 coplanar pairs of squares (Fig. 131).

Part II Compounds of Cubes

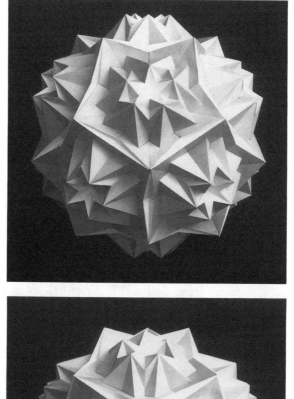
a

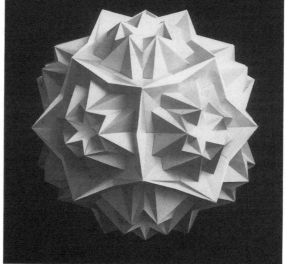
b

Figure 130. 20| $A_5 \times I$ / $C_3 \times I$ |31°02'41".91: (a) along a twofold axis; (b) along a threefold axis; (c) along a fivefold axis; (d) closeup along a fivefold axis

Chapter 5 Stability of Subcompounds

c

d

The rotational angles of the two subcompounds $4|(A_4 \times I \,/\, C_3 \times I \,|\, \mu)$ are

$$\mu_a = 23°25'51''.28$$
$$\mu_b = 23°25'51''.28 + 44°28'39''.04 = 67°54'30''.33$$

Hence,

$$20|\, A_5 \times I \,/\, C_3 \times I \,|\, 31°02'41''.91$$
$$= 5 \times 4|\, A_4 \times I \,/\, C_3 \times I \,|\, 23°25'51''.28$$
$$= 5 \times 4|\, A_4 \times I \,/\, C_3 \times I \,|\, 52°05'29''.67$$

Note that the first subcompound, which is illustrated in Fig. 132, is very similar to $4|\, A_4 \times I \,/\, C_3 \times I \,|\, 22°14'19''.52$ in $10|\, A_5 \times I \,/\, D_3 \times I \,|\, B$ (see Section 2.4b).

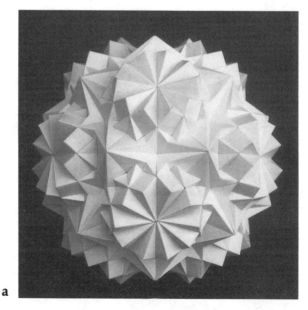

a

Figure 131. $20|\, A_5 \times I \,/\, C_3 \times I \,|\, 23°25'51''.28$: (a) along a twofold axis; (b) along a threefold axis; (c) along a fivefold axis

Chapter 5 Stability of Subcompounds

b

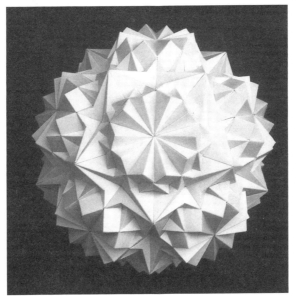

c

Figure 132. General view of $4| A_4 \times I / C_3 \times I | 23°25'51''.28$

c. $20| A_5 \times I / C_3 \times I = 5 \times 4| A_4 \times I / C_3 \times I$ (2×), except for $\mu_5 = 15°31'20''.96$ where the disjunctive subcompounds $4| (A_4 \times I / C_3 \times I)$ are

$4| (A_4 \times I / C_3 \times I)|15°31'20''.96$
$\qquad = 4| A_4 \times I / C_3 \times I |15°31'20''.96$ (a)

$4| (A_4 \times I / C_3 \times I)|60°$
$\qquad = 4| S_4 \times I / D_3 \times I$ (b)

Version (b) has a superior stability $S_4 \times I / D_3 \times I$ and is rigid. Hence,

$20| A_5 \times I / C_3 \times I |15°31'20''.96 = 5 \times 4| S_4 \times I / D_3 \times I$

d. $20| A_5 \times I / C_3 \times I = 10 \times 2| D_3 \times I / C_3 \times I$, except for $\mu_6 = 7°45'40''.48$ where the disjunctive subcompound $2| (D_3 \times I / C_3 \times I)$ has a superior stability $D_6 \times I / D_3 \times I$ and becomes rigid (see Chapter 4, Section 4.2.5, $n = 1$; corresponding to the end position). Hence,

$20| A_5 \times I / C_3 \times I |7°45'40''.48 = 10 \times 2| D_6 \times I / D_3 \times I$

Chapter 5 Stability of Subcompounds

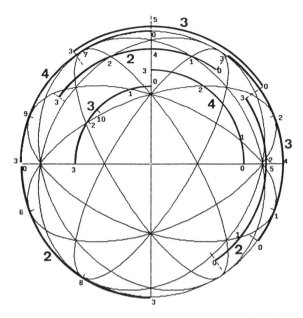

Figure 133. Special positions of $30A \mid A_5 \times I \mid C_2 \times I$ in Fig. 92

2. $30A \mid A_5 \times I \mid C_2 \times I$

Seven special positions are found in Fig. 133.

a. $30A \mid A_5 \times I \mid C_2 \times I = 6 \times 5A \mid D_5 \times I \mid C_2 \times I\,(2\times)$, except for $\mu_7 = 31°43'02''.91$ and $\mu_8 = 58°16'57''.09$ where alternatively one of both disjunctive subcompounds $5A \mid (D_5 \times I \mid C_2 \times I)$ has a superior stability, respectively, $D_{20} \times I \mid D_4 \times I$ and $D_{10} \times I \mid D_2 \times I$, and becomes rigid (see Chapter 4, Section 4.2.7, $n = 5$; respectively corresponding to the initial and the end positions). Hence,

$30A \mid A_5 \times I \mid C_2 \times I \mid 31°43'02''.91 = 6 \times 5 \mid D_{20} \times I \mid D_4 \times I$
$30A \mid A_5 \times I \mid C_2 \times I \mid 58°16'57''.09 = 6 \times 5 \mid D_{10} \times I \mid D_2 \times I$

b. $30A \mid A_5 \times I \mid C_2 \times I = 10 \times 3A \mid D_3 \times I \mid C_2 \times I\,(2\times)$, except for $\mu_6 = 20°54'18''.56$ and $\mu_9 = 69°05'41''.44$ where alternatively one of both disjunctive subcompounds $3A \mid (D_3 \times I \mid C_2 \times I)$ has a superior stability, respectively,

$D_6 \times I / D_2 \times I$ and $D_{12} \times I / D_4 \times I$, and becomes rigid (see Chapter 4, Section 4.2.7, $n = 3$; respectively corresponding to the end and the initial positions). Hence,

$30A | A_5 \times I / C_2 \times I | 20°54'18''.56 = 10 \times 3 | D_6 \times I / D_2 \times I$
$30A | A_5 \times I / C_2 \times I | 69°05'41''.44 = 10 \times 3 | D_{12} \times I / D_4 \times I$

c. $30A | A_5 \times I / C_2 \times I = 15 \times 2 | D_2 \times I / C_2 \times I$, except for three positions where the disjunctive subcompound $2 | (D_2 \times I / C_2 \times I)$ has a superior stability (see Chapter 4, Section 4.2.6; $n = 2$). For $\mu_4 = 35°15'51''.80$ and $\mu_5 = 54°44'08''.20$, the stability is $D_6 \times I / D_3 \times I$ (each corresponding to a position of additional coincidence $3 \mapsto 2$). For $\mu_{10} = 45°$, the stability is $D_4 \times I / D_2 \times I$ (corresponding to the end position). In each of these versions, the subcompound becomes rigid. Hence,

$30A | A_5 \times I / C_2 \times I | 35°15'51''.80 = 15 \times 2 | D_6 \times I / D_3 \times I$
$30A | A_5 \times I / C_2 \times I | 54°44'08''.20 = 15 \times 2 | D_6 \times I / D_3 \times I$
$30A | A_5 \times I / C_2 \times I | 45° \quad\quad = 15 \times 2 | D_4 \times I / D_2 \times I$

3. $30B | A_5 \times I / C_2 \times I$

Five special positions are found in Fig. 134.

a. $30B | A_5 \times I / C_2 \times I = 6 \times 5B | D_5 \times I / C_2 \times I (2 \times)$, except for $\mu_2 = 13°16'57''.09$ and $\mu_5 = 31°43'02''.91$ where alternatively one of both disjunctive subcompounds $5B | (D_5 \times I / D_2 \times I)$ has a superior stability, respectively, $D_{10} \times I / D_2 \times I$ and $D_{20} \times I / D_4 \times I$ and becomes rigid (see Chapter 4, Section 4.2.8, $n = 5$; respectively corresponding to the end and the initial positions). Hence,

$30B | A_5 \times I / C_2 \times I | 13°16'57''.09 = 6 \times 5 | D_{10} \times I / D_2 \times I$
$30B | A_5 \times I / C_2 \times I | 31°43'02''.91 = 6 \times 5 | D_{20} \times I / D_4 \times I$

b. $30B | A_5 \times I / C_2 \times I = 10 \times 3B | D_3 \times I / C_2 \times I (2 \times)$, except for $\mu_3 = 20°54'18''.56$ and $\mu_4 = 24°05'41''.44$ where alternatively one of both disjunctive subcompounds $3B |$

Chapter 5 Stability of Subcompounds

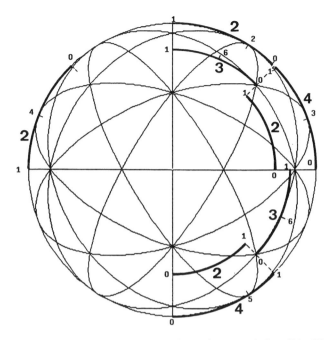

Figure 134. Special positions of $30B| A_5 \times I / C_2 \times I$ in Fig. 93

($D_3 \times I / C_2 \times I$) has a superior stability, respectively $D_{12} \times I / D_4 \times I$ and $D_6 \times I / D_2 \times I$ and becomes rigid (see Chapter 4, Section 4.2.8, $n = 3$; respectively corresponding to the initial and the end positions). Hence,

$30B| A_5 \times I / C_2 \times I |20°54'18''.56 = 10 \times 3| D_{12} \times I / D_4 \times I$

$30B| A_5 \times I / C_2 \times I |24°05'41''.44 = 10 \times 3| D_6 \times I / D_2 \times I$

c. $30B| A_5 \times I / C_2 \times I = 15 \times 2| D_4 \times I / C_4 \times I$, except for $\mu_6 = 22°30'$ where the disjunctive subcompound $2| (D_4 \times I / C_4 \times I)$ has a superior stability $D_8 \times I / D_4 \times I$ and becomes rigid (see Chapter 4, Section 4.2.4, $n = 1$; corresponding to the end position). Hence,

$30B| A_5 \times I / C_2 \times I |22°30' = 15 \times 2| D_8 \times I / D_4 \times I$

4. Special Versions with Spherical Freedom

Because the stabilizer is $E \times I$, all orientative subcompounds are disjunctive and have spherical freedom. For a subgroup $C_f \times I$, special positions are found when an axis of the descriptive cube is coincidental with, or perpendicular to, the sole axis (see Chapter 4, Section 3.3) where the subcompound version has a superior stability. Some examples are (with coincidence of axes)

$$12 | A_4 \times I / E \times I = 3 \times 4 | D_6 \times I / C_3 \times I \quad (3 \mapsto 2)$$
$$24 | S_4 \times I / E \times I = 4 \times 6 | D_6 \times I / D_1 \times I \quad (2 \mapsto 3)$$
$$= 3 \times 8 | D_{12} \times I / C_3 \times I \quad (3 \mapsto 4)$$
$$= 4 \times 6 | D_{12} \times I / C_4 \times I \quad (4 \mapsto 3)$$
$$60 | A_5 \times I / E \times I = 6 \times 10 | D_{10} \times I / C_2 \times I \quad (2 \mapsto 5)$$
$$= 6 \times 10 | D_{15} \times I / C_3 \times I \quad (3 \mapsto 5)$$
$$= 6 \times 10 | D_{20} \times I / C_4 \times I \quad (4 \mapsto 5)$$
$$= 10 \times 6 | D_6 \times I / D_1 \times I \quad (2 \mapsto 3)$$
$$= 10 \times 6 | D_{12} \times I / C_4 \times I \quad (4 \mapsto 3)$$

etc.

Chapter 6

Higher Descriptives

Our last subject to be discussed is the orbit system of a higher body whose system on level (2) is composed of congruent, centered cubes. Let p (and specifically p_i) denote centered positions of a cube for an *I*-group **G**.

1. Subcompound

A subcompound in a compound of cubes is a space-(2) body whose components are congruent, centered cubes. The orbit of the subcompound determines a space-(3) system, on level (2) being the compound itself (see Chapter 2, Section 2).

2. Decomposition Sequence

When applying the construction of Chapter 2, Section 6.2 for a centered cube as the space-(1) body p, a compound of cubes can be seen as a complex combination of different orbits of higher descriptives whose systems on level-(2) are always identical with the compound itself.

3. Compound Sum

Let $\mathbf{p} \in V^{(2)}$ be composed of cubes p. The orbit $\mathbf{G}(\mathbf{p})$ is then a level-(3) body, whose system on level-(2) is the sum of the compounds described by the components of $\mathbf{p}$:

$$\bigcup \{\mathbf{G}(p) \mid p \in \mathbf{p}\}$$

Some examples follow.

3.1. $10|\ A_5 \times I\ /\ D_3 \times I|\ A\ +\ 10|\ A_5 \times I\ /\ D_3 \times I|\ B$

This example was already seen in Chapter 5, Section 2.4: The sum of both versions of $10|\ A_5 \times I\ /\ D_3 \times I$ is the orbit system on level (2) of a space-(2) descriptive $2|\ D_6 \times I\ /\ D_3 \times I$ for $A_5 \times I$. This body is composed of 20 centered cubes, having overall symmetry for $A_5 \times I$.

3.2. One Extra Cube

A compound for an I-subgroup $\mathbf{F} \subset S_4 \times I$ may always be increased to an arrangement

$$\mathbf{F}(p_1)\ \cup\ 1|\ S_4 \times I\ /\ S_4 \times I$$

This arrangement is the orbit system on level (2) of a body $\mathbf{p} = \{p_1, p_2\}$ for $\mathbf{F}$, whose second component p_2 has a stability $S_4 \times I\ /\ S_4 \times I$.

Two examples are illustrated.

1. *The "Theosophical" Compound*

This arrangement of five cubes is the sum of $4|\ S_4 \times I\ /\ D_3 \times I$ with a cube (see also the Historical Appendix). The two components of $\mathbf{p}$ have a stability $S_4 \times I\ /\ D_3 \times I$ and $S_4 \times I\ /\ S_4 \times I$, respectively (Fig. 135).

2. *Seven Cubes under $S_4 \times I$*

This arrangement of seven cubes is the sum of $6|\ S_4 \times I\ /\ C_4 \times I$, the only uniform compound of cubes with rotational freedom, and a single cube. The two components of $\mathbf{p}$ have a stability $S_4 \times I\ /\ C_4 \times I$, and $S_4 \times I\ /\ S_4 \times I$, respectively (Fig. 136).

3.3. Sum of Proper Compounds

Any two or more *proper* compounds for an I-group $\mathbf{G}$ can be combined into an arrangement with overall symmetry for $\mathbf{G}$.

Figure 135. $4 | S_4 \times I / D_3 \times I$ and the "Theosophical" compound: (a) along a twofold axis; (b) along a fourfold axis

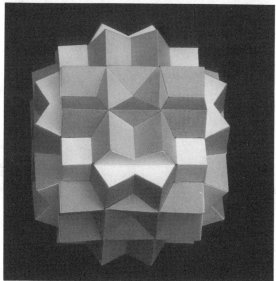

Figure 136. $6 \mid S_4 \times I \ / \ C_4 \times I \mid 15°$ and "seven" cubes

Chapter 6 Higher Descriptives

Figure 137. 15 + 5 cubes: (a) along a twofold axis; (b) along a fivefold axis

As an example, a sum

$$15 \mid A_5 \times I \mid D_2 \times I + 5 \mid A_5 \times I \mid A_4 \times I$$

is an arrangement of 20 centered cubes, with overall symmetry for $A_5 \times I$. This arrangement is the orbit system on level (2) for $A_5 \times I$, of a space-(2) body **p**, whose two components are positions with a stability $A_5 \times I \mid D_2 \times I$ and $A_5 \times I \mid A_4 \times I$, respectively (Fig. 137).

Chapter 7

Assembling Models

Three-dimensional images have always been a highly desirable source of inspiration since the previous century. Who has never seen one of these wooden "belle époque" 3D-viewers, in which pairs of pictures had to be inserted? Or colored printed figures—and even movies—that have to be viewed with red-and-green spectacles?

Unfortunately, the photographs of models shown here are actually projections of three-dimensional models. Seeing such a model in reality is still a different experience. When each cube in the model is made of a different color, the distinct cubes can be recognized easily. Therefore, an explanation of how a number of models can be assembled by yourself is presented in this chapter. Not only will this add to a better understanding of the treated theory, it may also prove to be a source of great fun; the assembly of such a model is similar to that of a jigsaw puzzle—but in space.

In the added patterns, the parts are indicated in bold lines and the edges are numbered. All you have to do is join two pieces with the same number until the entire puzzle is assembled; your three-dimensional model is finished!

1. Assembly Principles

The pieces must be made from a soft cardboard sheet, something like a stiff paper. Not too thick; the sheet should be easily foldable. A sheet similar to that used for namecards would be ideal. The parts have to be cut out and provided with folded tabs, which are to be glued together with white water-solvable wood glue. Your first model should be an easy one, with a relatively small amount of pieces. If you want to build a colored model, I would advise starting with a rough work model in one color. By assembling this one, you will quickly develop a technique, and, then, a final colored model will be fairly easy to assemble, as you can identify the pieces of each cube now on your work model.

A more difficult, but exciting assembling principle is to start later from a subcompound, which you assemble first and on top of which you continue the actual compound model. This way is to be chosen only if you have become well acquainted with the assembly process, as is explained later. From the patterns, you will have to isolate the shape of the parts in the subcompound. Although this technique takes more skill, it is my favorite one: for, first, the model is far more rigid because it is solid inside; second, it will provide the ideal illustration for a subcompound.

In Fig. 138, such a sequence is shown for $6 | A_4 \times I / C_2 \times I | 22°30'$, starting from the disjunctive dihedral subcompound $2 | D_2 \times I / C_2 \times I | 22°30'$ (Fig. 138).

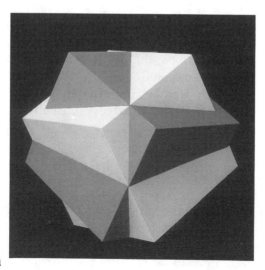

a

Chapter 7 Assembling Models

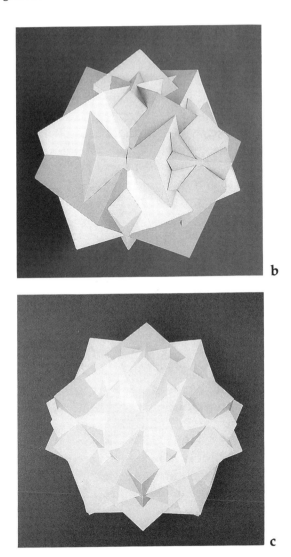

b

c

Figure 138. (a) Model of disjunctive subcompound $2|\ D_2 \times I\ /\ C_2 \times I\ |22°30'$; (b) assembly on top of the subcompound model; (c) finished model of $6|\ A_4 \times I\ /\ C_2 \times I\ |22°30'$

2. Preparing the Puzzle Pieces

First, take a photocopy of the patterns for the model you have chosen to assemble. The number of patterns vary from one to two, according to the compound. Only one pattern occurs when the dual—a compound of octahedra—is uniform. Hence, all vertices are there alike, which is dualized by the likeliness of all faces for the cube compound. From compounds of eight cubes or more, it is advisable to take an enlarged copy on A3 format, which is the original format on which the patterns were drawn and from which the pictured models were assembled for this book.

In addition to sheets of white and colored cardboard paper, the following tools are now needed:

- ◊ a sharply pointed object (like a needle in a holder, or a compass point)
- ◊ a bluntly pointed object (like a stencil pen)
- ◊ a small ruler
- ◊ a pair of scissors
- ◊ a small jar of white glue, with a fine, pointed opening

With the needle, the vertices of a piece are "pricked" into the white cardboard sheet, directly from your copied pattern. Then, use the blunt point and the ruler to press/draw firmly the line segments of the piece. Cut the piece out, providing tabs. This may be done roughly because the tabs do not have to match, as they are on the inside of the future model (Fig. 139). Because to the lines marked by the blunt point, the tabs are easily folded down. Now, the piece is ready to be glued onto the next one.

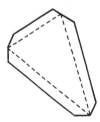

Figure 139. Triangular part with foldable tabs

3. Assembling the Pieces

All pieces on the patterns are numbered, either by lowercase letters or by capital letters. Line segments of pieces have to be glued together along the tabs, folded inward, and following identical letters. Capital letters indicate line segments in symmetry planes of the compound. Hence, two enantiomorphous (reflected) such pieces have to be glued together along that segment. Therefore, a number of pieces will have to be prepared oppositely, which will become clear when the model is proceeding. Numbers ($n = 2, 3, 4$, or 5) indicate intersecting points of the face with an n-fold symmetry axis of the compound.

4. Colored Models

When the work model is finished in white cardboard, you will easily identify the constituent cubes in the compound. Then, you may choose a number of colored sheets, equal to that of the constituents. Start all over now, using colored pieces for the correspondingly colored constituent.

5. Patterns

The following patterns are provided, good for the assembling of 10 models. The sequence is based on the degree of difficulty. Hence, the first model is the easiest, and the last one the most tedious.

5.1. Compound Sum of Seven Cubes under $S_4 \times I$

See Chapter 6, Section 3.2.2 and Fig. 136. Version 6 | $S_4 \times I$ / $C_4 \times I$ | 15° is ideal for combining with the extra cube. About each fourfold axis, twice three squares are described in a regular dodecagon {12}, turning this sum quite attractive (Fig. 140).

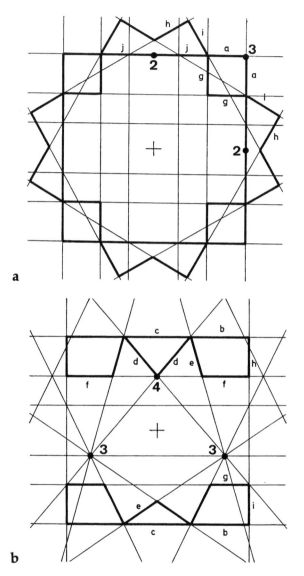

Figure 140. Patterns for $6|\,S_4\times I\,/\,C_4\times I\,|15° + 1|\,S_4\times I\,/\,S_4\times I$: (a) 6 faces; (b) 24 faces

5.2. The Classic Compound of Five Cubes

See Chapter 4, Section 4.5.1 and Fig. 108. $5 \mid A_5 \times I \mathbin{/} A_4 \times I$ is the dual of a uniform compound of five octahedra [9]. (See Fig. 141.)

Four cubes of this compound constitute $4 \mid A_4 \times I \mathbin{/} C_3 \times I$ $\mid 44°28'39''.04$ (Fig. 142). See Chapter 5, Section 3.1.1 and Fig. 120.

As an exercise, this may be your start to building a compound model on top of a subcompound. After you have assembled both models, observe the pattern on Fig. 141 to locate the pieces in Fig. 142. Then you may add the remaining pieces on top of your model of four cubes to obtain a second model of five cubes.

5.3. Tetrahedral Models of Four Cubes out of Ten

The two subcompound versions of $4 \mid A_4 \times I \mathbin{/} C_3 \times I$, in the two versions of $10 \mid A_5 \times I \mathbin{/} D_3 \times I$ follow next (see Chapter 5, Section 2.4 and Figs. 98–99, and the cover). Together with the previous special model of four cubes, these three models provide a nice illustration of the rotational freedom of a compound. The duals are versions of a uniform compound of octahedra, with rotational freedom [9]. See Figs. 143 and 144.

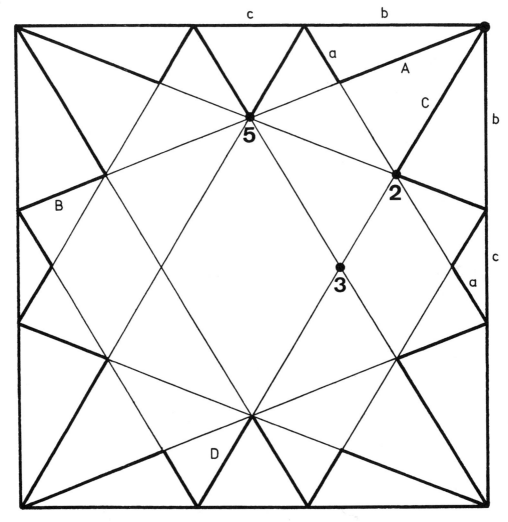

Figure 141. Pattern for the 30 faces of $5 \mid A_5 \times I / A_4 \times I$

Chapter 7 Assembling Models

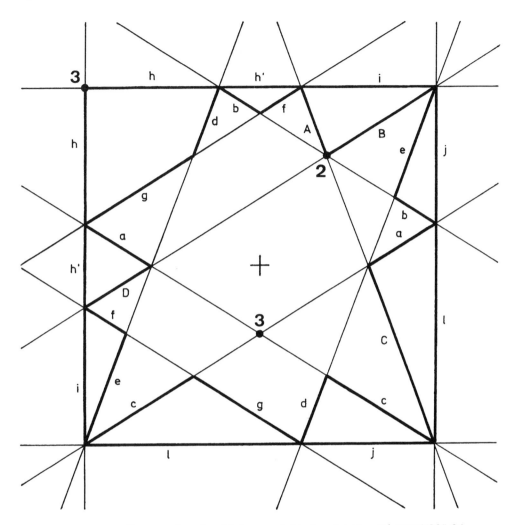

Figure 142. Pattern for the 24 faces of $4\mid A_4 \times I \mathbin{/} C_3 \times I \mid 44°28'39''.04$

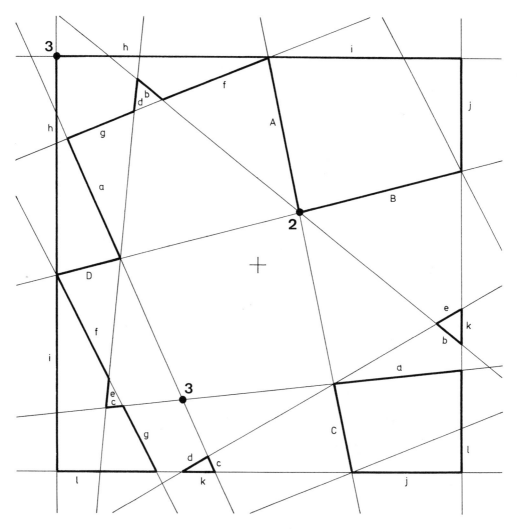

Figure 143. Pattern for the 24 faces of $4 \mid A_4 \times I \mid C_3 \times I \mid 22°14'19''.52$

Chapter 7 Assembling Models 211

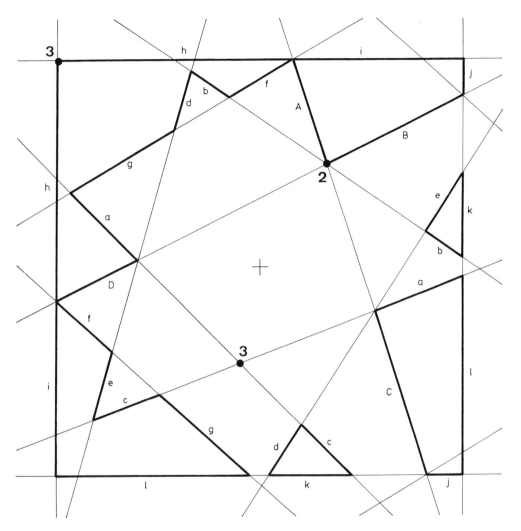

Figure 144. Pattern for the 24 faces of $4|\ A_4 \times I\ /\ C_3 \times I\ |37°45'40''.48$

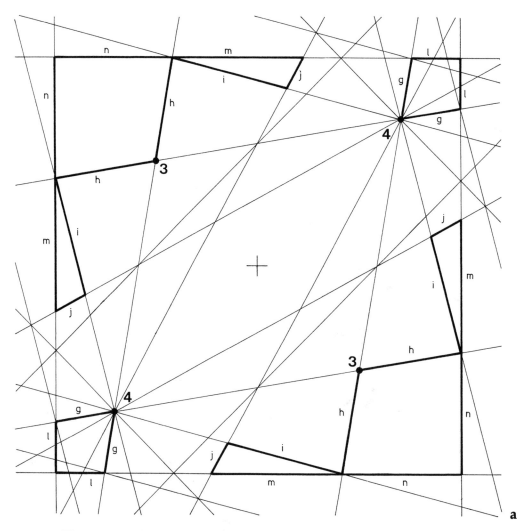

Figure 145. Patterns for $6 | S_4 \times I / D_2 \times I$: (a) 12 faces; (b) 24 faces

5.4. The Rigid Octahedral Model of Six Cubes

See Chapter 4, Section 4.4.4 and Fig. 104. Pattern distribution: a (12 faces); b (24 faces). See Fig. 145.

Chapter 7 Assembling Models

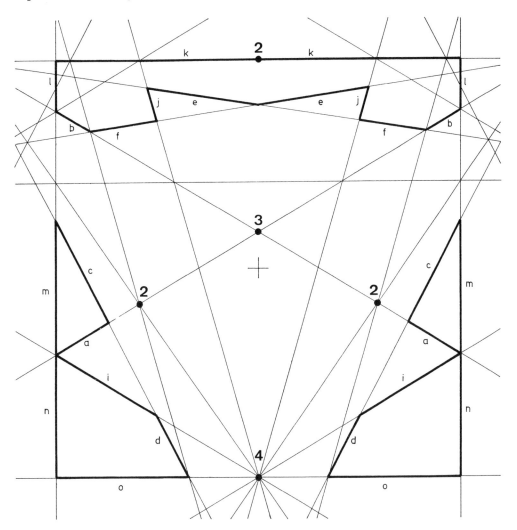

Figure 145b.

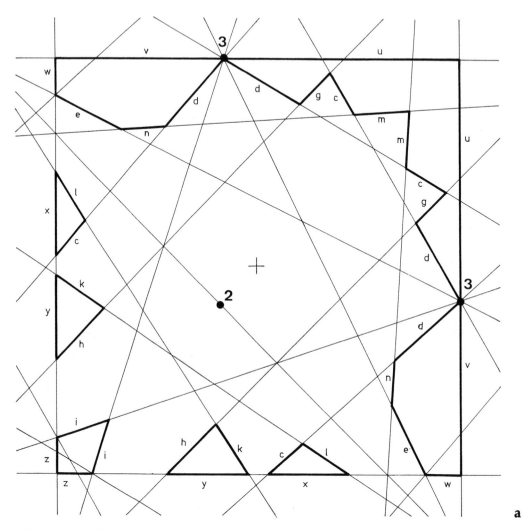

Figure 146. Patterns for $6 \mid A_4 \times I \mid C_2 \times I \mid 14°21'33''.24$: (a) 12 faces; (b) 24 faces

5.5. Tetrahedral Models of Six Cubes

Two versions are presented: $6 \mid A_4 \times I \mid C_2 \times I \mid 14°21'33''.24$ and $4 \mid A_4 \times I \mid C_2 \times I \mid 22°30'$ (see Chapter 4, Section 4.3.2 and Figs. 100–

Chapter 7 Assembling Models

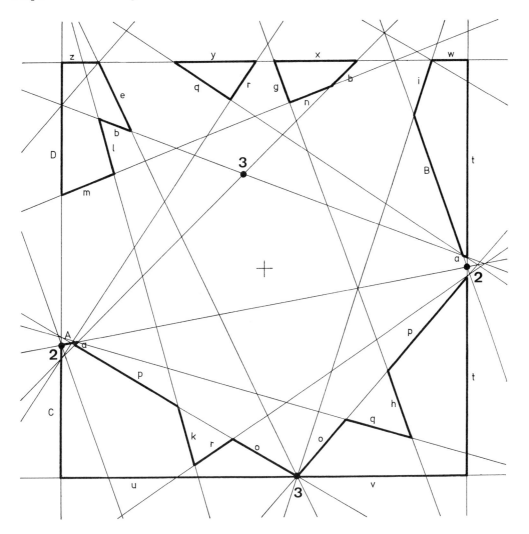

Figure 146b.

101). As for the previous model, pattern a is good for 12 faces, and pattern b for 24 faces. These three models of six all illustrate the rotational freedom of the orientative compound $(6 \mid A_4 \times I \mid C_2 \times I)$. See Figs. 146 and 147.

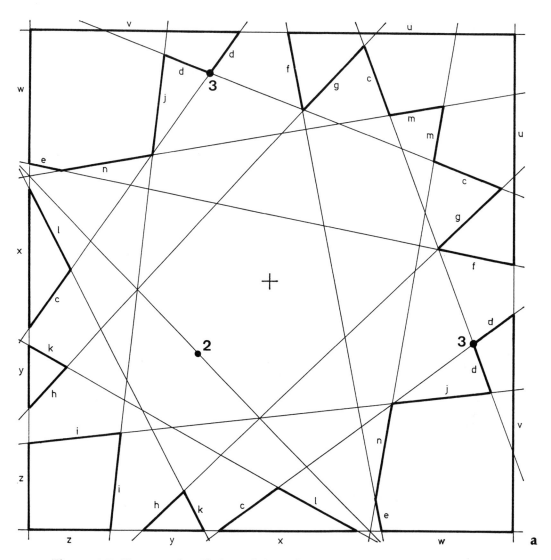

Figure 147. Patterns for $6 | A_4 \times I / C_2 \times I | 22°30'$: (a) 12 faces; (b) 24 faces

Chapter 7 Assembling Models

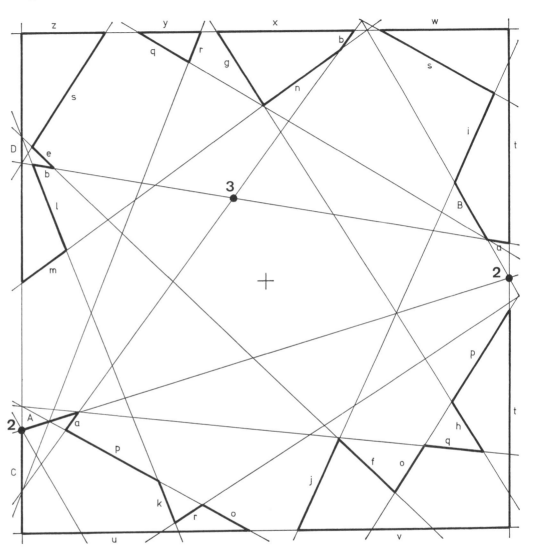

Figure 147b.

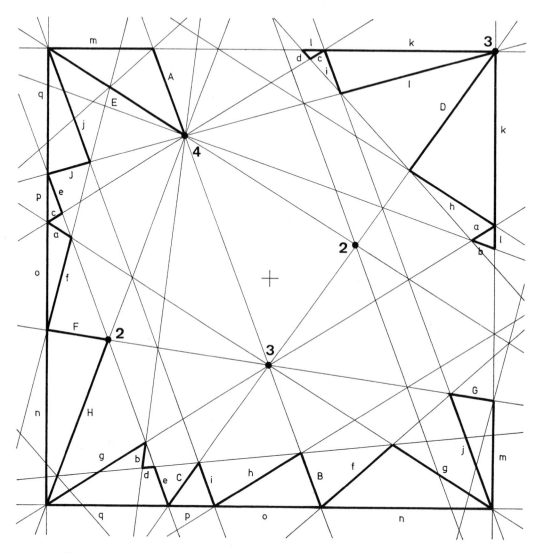

Figure 148. Pattern for the 48 faces of $8| S_4 \times I / C_3 \times I |44°28'39''.04$

5.6. Eight Cubes under $S_4 \times I$

See Chapter 5, Section 3.2.2a and Fig. 106. $8| S_4 \times I / C_3 \times I |44°28'39''.04$ contains two special tetrahedral subcompounds of four cubes (Section 5.2) and has two constituents per vertex. Its dual compound of eight octahedra is uniform [9], and the pattern is good for 48 faces.

5.7. Twenty Cubes under $A_5 \times I$

See Chapter 5, Section 3.3.1a and Fig. 130. 20| $A_5 \times I$ / $C_3 \times I$ |31°02′41″.91 contains five special tetrahedral subcompounds of four cubes (Section 5.2) and has two constituents per vertex. Its dual compound of 20 octahedra is uniform [9], and the pattern is good for 120 faces (Fig. 149).

Figure 150 illustrates the drawn stellation pattern of the compound core which had to be analyzed for the visible parts.

220   Part II Compounds of Cubes

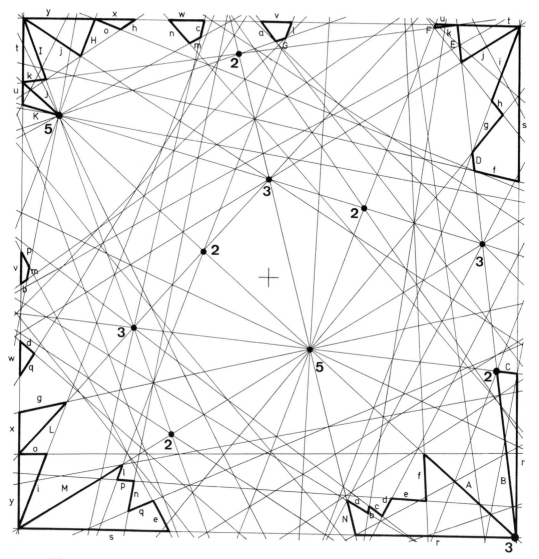

Figure 149. Pattern for the 120 faces of $20| A_5 \times I / C_3 \times I |31°02'41''.91$

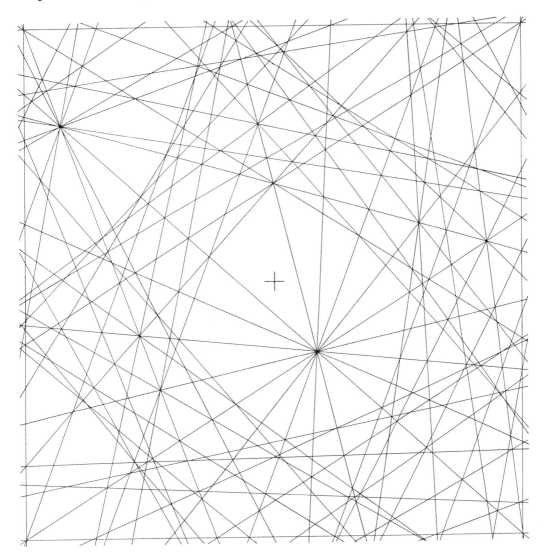

Figure 150. Stellation pattern before parts analysis

Appendix

Historical Survey

It is remarkable how events in life are greatly determined by our relations with people. A small gesture may alter the course of a lifetime. As an example, here is a story of what happened to me when I was a boy in high school, still living with my parents in a suburb of the city of Antwerp.

In 1966, our downstairs neighbor, J. Geerts, had noted my special interest in geometry and had offered me, for my 17th birthday, a magnificent old book from 1915 from his extensive esoterical library. The book, bound with linen, and printed on hand-extended paper, was called in Dutch (= Flemish) *Meetkunde en Mystiek* (*Geometry and Mystics*) and is presently still in my possession. The author, Dr. H. Naber, had based his book on his lectures held for the Theosophical Society in the Netherlands on the subject of the Golden Section in Platonic solids and on the measures of the Great Pyramid at Giza, Egypt. Naber knew that the key to the geometry of the Great Pyramid was the Golden Number, which he illustrated with some geometrical examples. Also, a remarkable drawing of a description of the five Platonic solids in one another was shown, based on publications by J. Kepler in the early 17th century. Kepler had observed that the five Platonic solids could be described in one another, following the sequence: tetrahedron—cube—dodecahedron—icosahedron—octahedron—tetrahedron, and from this point, starting again. Moreover, Naber explained that each inscription *revealed the Golden Number,* in a particular way. The book made a deep impression and opened a way for further investigation.

In 1972, I graduated in mathematics, together with my roommate K. Konings, with a jointly written script on geometrical topology. Konings specialized further in computer science, whereas I was looking for fields in geometry for specialization. The two main subjects of Naber's book still intrigued me, and in order to work with three-dimensional models, I felt it necessary to develop some techniques of model construction. The first model I ever built was the impression of the Kepler–Naber sequence of Platonic solids made of rods and balls (Fig. 151).

In this study model, I noted that a compound of cubes was easily obtained by five such cubes that could be inscribed in the dodecahedron, being the standard compound of five cubes. A first publication of this compound, which is here denoted by $5 | A_5 \times I / A_4 \times I$, is by Hess, back in 1876 [6], a second by Klein in 1884 [7], and a third by Brückner in 1990 [2]. Brückner introduced two more compounds of cubes, here denoted by $2 | D_6 \times I / D_3 \times I$ and $3 | S_4 \times I / D_4 \times I$. Each of the three compounds was illustrated by a picture of a cardboard model (in 1900!). Gradually, I was getting more involved with the Golden Number in polyhedral shapes, and constructed a collection of large structures in fine materials with light within, expressing geometrical properties.

One year later, in 1973, H.S.M. Coxeter introduced a number of definitions for compounds [5]:

> A *compound* is a set of equal regular polyhedra with a common center. The compound is *vertex-regular* if the vertices of its components are together the vertices of a single regular polyhedron, and *face-regular* if the face-planes of its components are the face-planes of a single regular polyhedron.

Five regular compounds are, thus, established in the book.

In 1974, W.G. Harman, of Camberley, England, wrote a remarkable manuscript on the classification of Polyhedral Compounds. This paper was never published, but a limited amount of copies was distributed among friends, and I obtained such a copy through M.G. Fleurent of the Abbey of Averbode, Belgium. Harman described in his 82-page document the construction of 72 symmetric compounds of Platonic solids, of which a number was illustrated by his own drawings. However, the

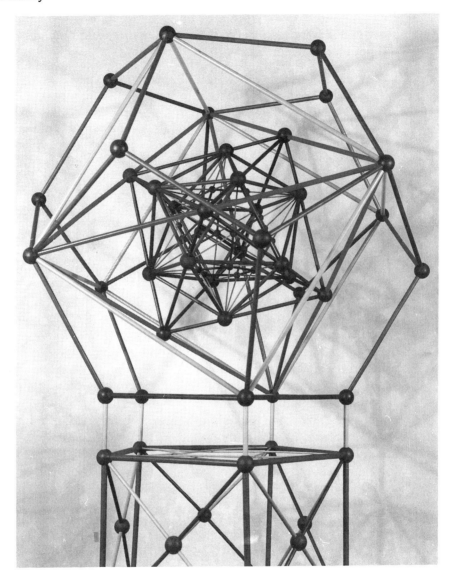

Figure 151. Model of the Platonic solids, described in one another (2.30 m high). The model fell down and broke into pieces when my cat had climbed in it.

descriptives were restricted to space-(1) bodies, and only rigid compounds were considered:

> A *regular compound* is a compound in which all components are similarly situated with respect to the overall symmetry of the compound, and the orientations of the components are fully determined by the overall symmetry of the compound. . . . The first part of the definition is designed to exclude what might be called *improper compounds,* i.e. compounds formed by the "addition" of two or more compounds having identical components and overall symmetry. . . . The second part of the definition excludes compounds which can be *continuously deformed,* by varying the orientations of the components, without affecting the overall symmetry of the compound.

Harman meant the exclusion of compound sums [i.e., compounds obtained by space-(2) descriptives] and compounds with central or rotational freedom. The treatment was rather chaotic, yet seven new rigid cube compounds were derived, which are $n \mid D_{4n} \times I / D_4 \times I$, $n \mid D_{3n} \times I / D_3 \times I$, $n \mid D_{2n} \times I / D_2 \times I$, $6 \mid S_4 \times I / D_2 \times I$, $10 \mid A_5 \times I / D_3 \times I$ (two versions), and $15 \mid A_5 \times I / D_2 \times I$ in my notation system.

In 1975, my geometrical objects started to attract the attention of the art circuit. A first exhibit was offered; this directed my entire energy for a few years to extending the collection. In that year, J. Skilling, of Cambridge, England, published a computer analysis which exhausted the complete set of *Uniform Polyhedra* [8]. A year later, a sequel followed, listing now the *Uniform Compounds* of these [9]. In this article, the idea of a "Uniform Compound of Uniform Polyhedra" was introduced, accordingly to the definition of a Uniform Polyhedron:

> A *Uniform Compound* is a three-dimensional combination of uniform polyhedra whose edge-lengths are all equal and whose relative position is such that the symmetry group of the combination is transitive on the set of all vertices of all the polyhedra.

Also the terms "constituent" and "rotational freedom" were introduced in the article, which I have accordingly taken over, and 75 Uniform Compounds were classified, among which 3 consisted of cubes: $3 | S_4 \times I / D_4 \times I$, $6 | S_4 \times I / C_4 \times I$, and $5 | A_5 \times I / A_4 \times I$. Of these, the compound $6 | S_4 \times I / C_4 \times I$ with rotational freedom was new. A first illustration, though, appears here in Chapter 4, Fig. 105.

The year 1978 was very busy for me. First, I was forced to move twice—including my collection of Light Objects—which was a very energy-consuming, but rather wasteful pastime. In February, a friend, E. Vandekerckhove, found another old book of interest—remarkably again a Theosophical work—in some corner of his brother's esoterical bookstore here in Antwerp, which he gave me (and which, afterward in 1986, I offered to M.J. Wenninger, of Collegeville, Minnesota, on the occasion of his second visit). It concerned the transactions of a Theosophical Congress, held in London in 1905 [14]—21 years after the appearance of Klein's book. In its contents, an extensive and richly illustrated paper "Notes on the Fourth Dimension," written by a certain W.J.L. (no full name provided), is found. Herein, the author described—among four-dimensional polytope properties—a three-dimensional symmetric arrangement of five cubes as being a compound of cubes, distinct from "Klein's compound," and derived from an initial cube and four rotated positions of it:

> Take the cube as a basic one and let it have its four diagonals common respectively one to each of the other four cubes; next let each of the four subsdidiary rotate about the diagonal common to the basic cube from a position of congruence with the latter through an angle of $\pi/3$. . . . This arrangement of five cubes is, of course, totally distinct from the arrangement of five cubes mentioned by Professor Klein, which *each* cube of the five has its four diagonals common respectively one to each of the other four, and their vertex points define those of a regular Platonic dodecahedron. In the arrangement which I have been describing only *one* cube, the basic one, has all its four diagonals common, one to each of the other four; the subsidiary cubes have only *one* of their diagonals in common, viz., with the basic cube.

This passage opened a new realm of investigation. The "Theosophical" compound is actually a sum of $4 | S_4 \times I / D_3 \times I$ and

$1 | S_4 \times I / S_4 \times I$, or, as outlined in Chapter 6, is obtained as the orbit of a space-(2) body of two component cubes. Much later, I found that T. Bakos published a description and a drawing of $4 | S_4 \times I / D_3 \times I$ in 1959 [1], and he probably was unaware of this previously published "Theosophical" sum with a single cube. Bakos's compound was, in fact, constructed as the dual of a compound of four octahedra, resulting in the four rotated cubes of the initial one in the construction of the "theosophical" compound.

In the summer of 1978, I was offered a first international exhibit of Light Structures in London. For some reason, Vandekerckhove had moved temporarily into a spacious apartment in London and invited me, together with Konings and his wife H. Thiers (who is a professional photographer, acknowledged in the Introduction), to his new place. One day, on a side trip to Oxford, Konings found a copy of Wenninger's book *Uniform Polyhedra* in a bookstore and purchased it for me. Some of my artwork on exhibit in London was based on Polyhedral Kaleidoscopes, and Konings had recognized kaleidoscopic-looking drawings in the book. Soon after, I started a correspondence with Wenninger, and, through him, I came in contact with Fleurent, who lived just outside Antwerp, hitherto without my knowledge.

In August, still 1978, I finally went to Egypt, as my art was doing fine in London, and I could now afford a three-week side trip for combining some research and holidays. Words are not sufficient to describe the ecstasy one is overwhelmed with when seeing the Great Pyramid for the first time. I surveyed the measures outside and within the passages and chambers, and discovered the *fivefold* geometry of it, which is related to the icosahedron in a simple, but fascinating way, and which is afterward published as a contribution in an anthology on fivefold symmetry [11].

One year later, in 1979, I was invited by Wenninger to a Monastery, in the Bahamas where he was living. I stayed there for a month in a tropical summer. New horizons appeared as he improved my techniques of model-making and as I was introduced to more Polyhedral enthusiasts. I had just become an assistant to Prof. Dr. W. Dyck at the University of Antwerp, who provided me with the opportunity and the funds to finish a doctoral dissertation of expandable structures, based on the jitterbug transformation principle of R. Buckminster Fuller. J.D. Clinton of Hillside, New Jersey, was once an assistant of Fuller and had already published part of the listing of the convex transformations, back in 1972, in a NASA report. Wenninger and Clinton were also acquainted, and, having informed Clinton of my arrival in the Bahamas, I was invited also to Hillside. When I arrived in New York City in August and was taken on a city tour by Clinton, the architectural wonders gave me almost a similar impression as the Great Pyramid a year before. We visited Fuller's office in Philadelphia, Pennsylvania. I felt the majesty of this great late philosopher still vibrating in the dramatic ubiquitous testimonies of his diversative imagination.

Two years went by, and, in the summer of 1981, I again traveled with M.J. Wenninger in North America, visiting a number of people who shared our interest with polyhedral matters, like H.S.M. Coxeter in Toronto, Ontario, A.L. Loeb in Cambridge, Massachusetts, N. Johnson in Wheaton, Massachusetts, and G. Odom, in Poughkeepsie, New York.

Odom succeeded in suggesting a good mental image of one of the two versions of the icosahedral compound of 10 cubes. As was mentioned before, a first description of this compound type with its two distinct versions was outlined in the unpublished manuscript of Harman, but no illustration was added. Therefore, it was rather hard to imagine the exact shape of each version, since merely a theoretical explanation was given. But Odom surprised me with an independently developed theoretical construction principle of this fascinating compound. He explained a transformation of the uniform compound of five cubes, $5 | A_5 \times I / A_4 \times I$, which appeared to be somehow similar to the transformation in the "Theosophical" compound; namely, each of the 10 threefold axes of $A_5 \times I$ is shared with a pair of cubes. If the cubes of each pair rotate simultaneously about these axes in opposite senses toward each other, they will quickly coincide, thus forming altogether a compound of 10 cubes. This compound version is here denoted by $10 | A_5 \times I / D_3 \times I | B$, but, from that time, I often refer to it as the "Odom-version." For, Odom was the first to understand a relationship between $5 | A_5 \times I / A_4 \times I$ and $10 | A_5 \times I / D_3 \times I | B$. As is explained in Chapter 4, Section 3.2.2, item 5 and Section 4.5.4, and in Chapter 5, diagram 5c, both compounds are special versions of the orientative compound $(20 | A_5 \times I / C_3 \times I)$.

Having arrived now at $(20 | A_5 \times I / C_3 \times I)$, a first ever full version—actually a special version of $20 | A_5 \times I / C_3 \times I$ with double square faces (see Chapter 5, Section 3.3.1b)—was published in 1982 by Wenninger, who discovered its incidental appearance in one of Fleurent's patterns of nonconvex dual snub polyhedra, exclusively prepared for the book *Dual Models* [12]. The dual compound of this version, which is a Uniform Compound of octahedra, was published by Skilling in 1976 and was called there simply "20 octahedra," as a special case of the general compound of 20 octahedra with rotational freedom [9]. Skilling had listed the Uniform Compounds, mentioning also the number of constituents per vertex. And the compound of 20 octahedra has 1 constituent per vertex, whereas "20 octahedra" has 2. Hence, when dualizing "20 octahedra," a compound of 20 cubes composed of coplanar pairs of squares is obtained.

Later, in 1990, Wenninger succeeded in constructing models of the two earlier mentioned compounds without having vital geometrical patterns at his disposal: the "Theosophical" compound from the description in Vandekerckhove's book, and $10 | A_5 \times I / D_3 \times I | B$ from Odom's visualization. A photograph of each model was published in a paper dedicated to the Golden Number [13]. One of the readers, P. Messer of Mequon, Wisconsin, felt almost spontaneously that the pictured model of $10 | A_5 \times I / D_3 \times I | B$ seemed somehow distorted. As is explained in Chapter 4, Section 4.5.4, a threefold axis of the descriptive in the orientative compound $(20 | A_5 \times I / C_3 \times I | -22°14'19".52)$ makes an angle of $1°26'02".16$ with a twofold axis of $A_5 \times I$. Hence, near every twofold axis, two vertices of the compound make a central angle of $2°52'04".32$. When examining the picture

Historical Survey

of the model in Ref. 11, Messer noted that the pairs of vertices were constructed as coincidental with the twofold axes, reducing the number of vertices by 30 and turning the faces of the cubes slightly into diamonds. His geometrical pattern preparation proved excellent for a corrected model, exclusively reassembled by Wenninger for Chapter 4, Fig. 110.

In May 1991, Wenninger visited me again here in Belgium, and, together with Fleurent, we spent 3 days at the Abbey in Averbode, evaluating our separate contributive materials on compounds of cubes and those of everyone that we knew of in the past. Fleurent had already noted a special version of $8 | S_4 \times I / C_3 \times I$ with two constituents per vertex ($8 | S_4 \times I / C_3 \times I | 44°28'39''.04$, see Fig. 125) and was now wondering if a generalized theory would be appropriate for a total listing and discussion of the compounds of cubes and their special versions. Somehow, the hitherto diversative approach to this subject seemed to have definitely stagnated, and, at this point, it was commonly agreed that I, as a theoretical mathematician, should take over from the very beginning.

And so, the idea of a book was born.

References

1. Bakos, T., Octahedra inscribed in a cube, *Math. Gaz.* **XLIII,** 17–20 (1959).
2. Brückner, M., *Vielecke und Vielflache,* Teubner, Leipzig, 1900.
3. Coxeter, H.S.M., *Introduction to Geometry,* Wiley, New York, 1961.
4. Coxeter, H.S.M., *Regular Polytopes,* Dover, New York, 1973.
5. Coxeter, H.S.M., Longuet-Higgins, M.S., and Miller, J.C.P., Uniform polyhedra, *Phil. Trans. Royal Soc. London* **264A,** 401–450 (1954).
6. Hess, E., *Ueber die gleicheckigen und gleichflächigen Polyeder,* Marburg II.I. 1876.
7. Klein, F., *Lectures on the Icosahedron and the Solution of Equations of the 5th Degree,* Dover, New York, 1913. (Reprint of *Lectures on the Icosahedron,* 1884.)
8. Skilling, J., The complete set of Uniform Polyhedra, *Phil. Trans. Royal Soc. London* **278A,** 111–135 (1975).
9. Skilling, J., Uniform compounds of uniform polyhedra, *Math. Proc. Phil. Soc. Cambridge* **79,** 447–457 (1976).
10. Verheyen, H.F., The complete set of jitterbug transformers and the analysis of their motion, *Comp. Math. Appl.* **17,** 203–250 (1989); *Symmetry 2: Unifying Human Understanding,* I. Hargittai, Ed. Pergamon, Oxford, 1989.
11. Verheyen, H.F., *The Icosahedral Design of the Great Pyramid, Fivefold Symmetry,* I. Hargittai, Ed. World Scientific, Singapore, 1992.
12. Wenninger, M.J., *Dual Models,* Cambridge University Press, New York, 1983.

13. Wenninger, M.J., Polyhedra and the Golden Number, *Symmetry* **1**, 37–40 (1990).
14. W.J.L. Notes on the fourth dimension, *Transactions of the Second Annual Congress of the Federation of European Sections of the Theosophical Society held in London, July 6–10, 1905,* London, published for the Council of the Federation, 1907.

Illustratory Contributions

1. Manual Drawings

Author: 1, 2, 3, 4, 5, 6, 7, 8, 9, 10, 11, 12, 13, 14, 15, 16, 17, 18, 19, 21, 26, 38, 39, 40, 41, 42, 43, 44, 45, 46, 47, 48, 49, 50, 51, 52, 53, 54, 55, 56, 57, 58, 59, 60, 61, 62, 63, 64, 66, 67, 69, 70, 71, 72, 73, 74, 75, 76, 80, 81, 82, 83, 84*, 85*, 86*, 87*, 88*, 90*, 91*, 92*, 93*, 118*, 122*, 123*, 124*, 126*, 127*, 128*, 129*, 133*, 134*, 139.
(*): based on a computer graphic by M. Fleurent (see Chapter 2).
M. Fleurent: 140, 141, 142, 143, 144, 145, 146, 147, 148, 149, 150.

2. Computer Graphics

M. Fleurent: 65, 68, 71, 74.
J. Skilling: 77, 78.

3. Pattern Design (of Models for Part II)

M. Fleurent: Cover Model, 98, 99, 100, 101, 104, 105, 106, 107, 112, 113, 120, 130, 131, 132.
P. Messer: 109, 110, 111.

4. Model Assembly

Author: Cover Model, 20, 22, 23, 24, 25, 27, 28, 29, 30, 31, 32, 33, 34, 35, 36, 37, 95, 96, 97, 98, 120, 121, 132, 138, 151.
J. Wenninger: 94, 99, 100, 101, 102, 103, 104, 105, 106, 107, 108, 109, 110, 111, 112, 113, 130, 131, 135, 136, 137.

5. Photographs

Author: 20, 22, 23, 24, 25, 27, 28, 29, 30, 31, 32, 33, 34, 35, 36, 37, 94, 97, 103, 104b, 104c, 105, 106a, 106b, 107a, 107b, 108, 109a, 110c, 112a, 112c, 113, 130, 131, 135, 136, 137, 138a.

D. Bruno: 99, 100, 101, 102, 104a, 106c, 107c, 109b, 109c, 110a, 110b, 111, 112b.

H. Thiers: Cover Model, 95, 96, 98, 120, 121, 132, 138b, 151.

(Illustrations used more than once are mentioned only once.)